T0298431

Basic Vacuum Technology

Basic Vacuum Technology

Second Edition

A Chambers

University of York

R K Fitch

Walmley, Sutton Coldfield

and

B S Halliday

Birkenhead, Merseyside

Institute of Physics Publishing
Bristol and Philadelphia

British Library Cataloguing-in-Publication Data

A catalogue record for this book is available from the British Library.

ISBN 0 7503 0495 2

Library of Congress Cataloging-in-Publication Data are available

First edition published 1989
Corrected reprint 1995
Second edition 1998

Cover images courtesy of Pfeiffer Vacuum Ltd.

Published by Institute of Physics Publishing, wholly owned by The Institute of Physics, London

Institute of Physics Publishing, Dirac House, Temple Back, Bristol BS1 6BE, UK

US Office: Institute of Physics Publishing, The Public Ledger Building, Suite 1035, 150 South Independence Mall West, Philadelphia, PA 19106, USA

Typeset in Ami-Pro by Brenda Chambers

CONTENTS

3 Pumps 60

4 Measurement of pressure 86

5 Vacuum materials and components 115

Preface to the First Edition

In March 1986 the Vacuum Group of The Institute of Physics held its first two-day training course in 'Vacuum Technology' at the University of Strathclyde. The course was sponsored by the British Vacuum Council and the lectures and seminars were given by the present authors. Since then the course has been presented on four other occasions and the same tutors have also given in-house courses at industrial and research establishments. During this time it has become very clear that there is a need for an up-to-date book on this subject which would be suitable for engineers, technicians and postgraduate workers who need to acquire expertise in vacuum techniques. Furthermore it was believed that such a book should be sufficiently broad in content to include not only methods of production, measurement and system design and testing, but also describe the overall behaviour of gases at these low pressures. It was also recognised that if the book were not to be restricted to library shelves its format should not make it prohibitively expensive. In all of these respects the authors hope that a satisfactory compromise has been achieved.

A Chambers
R K Fitch
B S Halliday
June 1989

Preface to the Second Edition

Vacuum technology continues to make a vitally important contribution to industrial and scientific activity and the need for basic education in the subject is as strong as ever. In this second edition we have built upon our teaching experience with the first edition and taken the opportunity to modify the treatment of certain subjects and introduce new ones. The omission from the first edition of any discussion of the thermal conductivity and viscosity of gases has been rectified, and new sections have been included on thermal transpiration, molecular drag and the measurement of flow. The concepts of the continuum (viscous) and molecular states of gas and the Knudsen number have been introduced at an earlier stage. More worked examples and illustrative numerical examples are included in the first two chapters and we have attempted to update the coverage to reflect progress in the subject. A number of new appendices have been added.

A Chambers
February 1998

Acknowledgments

The British Vacuum Council and the Vacuum Group of the Institute of Physics under the chairmanship of Dr J S Colligon of the University of Salford were instrumental in launching the short course which was the origin of the first edition of this book. It is a pleasure to again record our thanks for that initial stimulus, and to the British Vacuum Council (BVC) for its continuing support of courses in the years following the publication of the first edition. The current chairman of the BVC, Dr R J Reid of CLRC Daresbury, has assisted in the development of the course in a number of ways, for which we are grateful.

One of the authors, A Chambers, has for a number of years enjoyed a stimulating involvement in educational activities at Edwards High Vacuum International and wishes to record thanks to several of the staff scientists there, in particular to Dr Alan Troup, Technical Director, Gordon Livesey and Malcolm Baker, who have increased his appreciation of the scope of the subject, and some of its subtleties. Dr Andrew Chew, now a member of that group and formerly a research student in the Department of Physics at the University of York, was a greatly valued collaborator in work on molecular drag. Stimulating discussions of gas physics continue with David York, currently a research student in the department and partly supported by the Edwards Company. Also at the University of York, Professor Jim Matthew gave considerable support and encouragement to vacuum teaching activities while Head of the Department of Physics. More recently, the course on which the book continues to be based has benefited significantly from his direct involvement. David Fellows of Arun Microelectronics Ltd has generously supported teaching and project work in the department in various ways, and is gratefully acknowledged.

Finally we record thanks to Kathryn Cantley at IOP Publishing, who has been most helpful and constructive in her suggestions for the preparation of this new edition.

A Chambers
R K Fitch
B S Halliday

February 1998

Introduction
R K Fitch and A Chambers

The vacuum environment plays a basic and indispensable role in present day technology. It has an extensive range of applications in industrial production, and in research and development laboratories, where it is used by engineers, scientists and technologists for a variety of purposes. The reasons for this range of applications will be discussed shortly. First it is necessary to comment on the units of pressure used to specify a vacuum, and to establish the range of vacua involved.

The standard international (SI) unit of pressure is the pascal (Pa) which is equivalent to a force of 1 newton per square metre, but it is still very common to use the millibar (mbar) which is not an SI unit. The millibar is used throughout this text. It is a convenient and acceptable unit because, in the majority of applications of vacuum technology, the concept of force per unit area is not relevant. We are more likely to be interested, once our purpose is defined, in the density of molecules in the working volume which contains the vacuum. Furthermore it can be assumed that for some time yet many laboratories will have equipment which is calibrated in other units, such as the torr. The relationship between these units is

$$100 \, \text{Pa} = 1 \, \text{mbar} = 0.76 \, \text{torr}$$

In fact in many applications we are often interested only in the order of magnitude of the pressure, and then we can reasonably ignore the difference between the mbar and the torr.

The range of pressures encountered in the applications of vacuum technology extends over more than fifteen orders of magnitude and it is useful to divide the total pressure range into five regions, namely:

1	'low vacuum'	-	atmospheric pressure to 1 mbar
2	'medium vacuum'	-	1 mbar to 10^{-3} mbar
3	'high vacuum'	-	10^{-3} mbar to 10^{-8} mbar
4	'ultra-high vacuum'	-	10^{-8} mbar to 10^{-12} mbar
5	'extreme high vacuum'	-	less than 10^{-12} mbar

These five divisions are somewhat arbitrary but in each case it is possible to relate them to different physical properties of the residual gas in the vacuum.

Low vacuum. In this region, in which the pressure is still a significant fraction of the atmospheric pressure, the main property is the force exerted due to the pressure difference between atmosphere and the vacuum. Thus it is used for mechanical handling, vacuum forming and vacuum brakes. It is also used in the degassing of fluids and for vacuum impregnation in certain electrical components where, for example, dissolved air is removed from a fluid to improve its electrical insulation.

Medium vacuum. The applications in this range are extensive and include processes such as vacuum drying and vacuum freeze drying for the food and pharmaceutical industries, and vacuum distillation for the chemical industry. In many of these processes an important factor which has to be considered is the vapour pressure of the fluid, frequently water. It is necessary that the pressure in the system should be less than the saturated vapour pressure of the fluid at the appropriate temperature. Thus in vacuum drying at room temperature the pressure must be less than about 1 mbar, whereas for vacuum freeze drying in the range -50 °C to -180 °C, the pressure must be in the region of 10^{-2} mbar. The effect of the vapours on the vacuum pump must also be taken into account.

High vacuum. The high vacuum region has very many applications which include the production of special materials for the metallurgical, electronics and aircraft industries and other processes such as electron beam welding. In TV, x-ray and gas discharge tubes, electron microscopy and particle accelerators it is necessary to use high vacuum. Perhaps the most important process in this pressure range is that of vacuum evaporation of thin films for lens blooming and the many aspects of semiconductor manufacture. Nearly all of these applications require that the 'mean free path' - the average distance travelled by gas molecules between collisions - be greater than the dimensions of the vacuum chamber. The approximate value of the mean free path in terms of the pressure p, measured in mbar, is given by the equation

$$\text{mean free path} = 0.007/p \text{ cm.}$$

Thus at a typical high vacuum of 10^{-5} mbar the mean free path is about 7 m which is greater than the dimensions of typical laboratory vacuum chambers.

Ultra-high vacuum. The pressure in the atmosphere is about 10^{-10} mbar at a height of 10^6 m so that some space simulators require ultra-high vacuum. On the other hand research in thermonuclear fusion uses ultra-high vacuum in order to achieve extremely high gas purities when ultra-pure gases are back-filled into vacuum chambers. In this pressure range the mean free path is very large and it is more important to consider the molecule - surface collisions than the molecule - molecule collisions. If we assume that at a pressure of p mbar all gas molecules arriving at a surface stick to it until one complete layer is formed, then the time t seconds to form such a monolayer on that surface is given approximately by the equation

$$t = 3 \times 10^{-6}/p.$$

Thus at 10^{-6} mbar this time is a few seconds whereas at 10^{-10} mbar it is several hours, making it possible to conduct measurements on atomically clean surfaces. Ultra-high vacuum is therefore an essential requirement in all studies of clean surfaces using for example the techniques of field-electron and field-ion microscopy, scanning tunneling microscopy, electron diffraction, Auger electron spectroscopy, secondary ion mass spectrometry and photoelectron spectroscopy. Ultra-high vacuum is also necessary in the new large (~km) gravity wave interferometers which are currently being developed at a number of laboratories, in order to have stable optical paths for the internal laser beams.

Extreme high vacuum. Since about the mid-eighties there has been an increasing demand to achieve pressures less than 10^{-12} mbar in a number of specialised applications. For example in anti-particle accumulators in storage rings pressures less than 10^{-12} mbar are required in order that the loss rate of particles by collision with gas molecules is not excessive. In the fabrication of some low dimensional structures by molecular beam epitaxy pressures of the order of 10^{-12} mbar or less are used to achieve ultra-clean growth conditions. This is necessary because the growth rate of the deposits is only about one atomic layer per second and purities of 1 part in 10^8 are required, so that contamination rates due to residual gas capture must be exceedingly small.

The range of applications of vacuum technology is thus extremely wide, and this book is an attempt to provide an up to date description of the relevant basic theory, methods of production and measurement, design and testing of vacuum, and of the materials and components involved. The plan of the book closely follows the programme of the course on which it is based. Chapter 1 is about gases in general with emphasis on their microscopic, molecular behaviour. Chapter 2 builds on this to discuss gases in vacuum systems, how vacua are specified, the origins of residual gases and flow into the pump. An attempt is made to present a strong descriptive picture of phenomena prior to the use of mathematics to make their description quantitative. Chapter 3 is concerned with the various pumps which produce vacua and chapter 4 with how vacua are measured in terms of total and partial pressures. Chapters 5 and 6 deal with practical matters of materials and their fabrication, components and cleaning and chapter 7 with the indispensable skills of leak detection. Finally chapter 8 presents some typical systems which exemplify the principles and practice dealt with in earlier chapters. Various matters whose lengthier discussion in the text would be inappropriate are dealt with in appendices. In addition to the literature references associated with each chapter, a further reading list is given at the end of the book.

1

Gases

A Chambers

Before we deal with ways in which gas can be removed from a vessel to create a vacuum environment, we need to understand something about how gases behave in themselves. And to do that, we need to consider what happens in a gas on a microscopic scale, and appreciate that it is the energetic and chaotic motion of its constituent molecules that fundamentally determines its behaviour.

1.1 MOLECULES

Just as an atom is the smallest unit of a chemical element, so a molecule, which consists of atoms in combination, is the smallest unit of a chemical compound. Thus atoms of the elements carbon and oxygen, represented by symbols C and O respectively, may combine to form molecules of carbon monoxide, formula CO, or carbon dioxide, CO_2, by the establishment of *bonds* between the atoms. Although the physical means exist to dismantle atoms and investigate their internal structure, there is relatively little we need to know about this atomic architecture for the purposes of vacuum practice. It is sufficient to consider its role in *bonding* and its modification to form *ions*. This information will be needed when we consider surface processes and pressure measurement for example. But for many purposes we may consider atoms and molecules to be simple spheres of a certain size and mass. Let us elaborate.

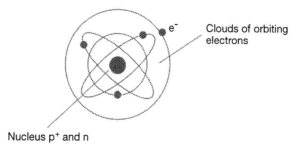

Figure 1.1 Representation of an atom.

As depicted in figure 1.1, an atom consists of a central positively charged *nucleus* made up of two types of particle, positively charged *protons* p^+ together with a comparable number of *neutrons* n. These are relatively heavy particles, with a mass of about 10^{-27} kg held together by the very strong nuclear force. Most of the atom's mass resides here in the nucleus. Outside the nucleus is a cloud of negatively charged orbiting *electrons* e^- so that the whole bears some resemblance to a planetary system. The electron, with a mass of about 10^{-30} kg, is much lighter than the particles in the nucleus and carries a charge equal in magnitude but opposite in sign to that of the proton. The electrical attraction between the nucleus and the electrons is what holds the atom together as an entity.

The number of protons in the nucleus is called the *atomic number* and is denoted by the symbol Z. An equal number Z of electrons orbit in the electron cloud so that the atom as a whole is electrically neutral. It is Z which gives the atom chemical identity. Thus $Z = 1$ for hydrogen, 2 for helium, . . ., 6 for carbon,. . ., 8 for oxygen. Of all the atoms, hydrogen is the simplest. Its nucleus consists of a single proton p+, around which a single electron e^- orbits. As remarked earlier, the atom for many purposes may simply be regarded as a sphere. The size of the sphere is roughly defined by the volume which the electron cloud occupies, figure 1.2.

Figure 1.2   Representing the atom as a sphere.

The *size* of atoms does vary with Z, but not in a dramatic way and for future illustrative purposes we will take a typical atom as having a diameter of 2×10^{-10} metre = 0.2 nm = 2 Å (Å = Angstrom unit).

1.2 BONDING

When molecules are formed by the establishment of bonds between atoms, it is the outer electrons of each atom which are involved in the bonding process. The bonding is essentially electrical in nature and arises from a net overall attraction between the positive nuclei and the negative electrons throughout the assembly. Bonding is to be explained only in terms of rather complicated laws, which fortunately need not concern us.

The intermingling of the atomic electron clouds which takes place in bonding is usually sufficiently intimate and the bonding sufficiently strong that in the case of simple molecules such as H_2, CH_4 etc a compact molecule results. Again for most of our purposes we can ignore the detailed structure and think of compact molecules as spherical, as in figure 1.3, with diameters of, typically, about 3 Å (0.3 nm).

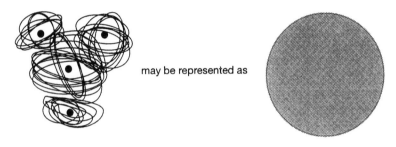

may be represented as

Figure 1.3   Representing the molecule as a sphere.

It is found that many of the common elemental gases, for example hydrogen, oxygen and nitrogen, exist most stably in molecular form as H_2, O_2 and N_2. Helium and argon gases however, whose atoms are chemically inert (i.e. not reactive), exist as single atoms.

1.3   IONS

When an atom or molecule loses an electron e⁻ it becomes electrically unbalanced and so carries a net positive charge. In this state the atom or molecule is said to be ionised and exists as an ion. H_2^+ and CO^+ are examples. Ions are created when, for example, fast moving free electrons collide with a target atom or molecule. It is the creation and manipulation of ions which is the basis of some total pressure gauges and all partial pressure analysers. We shall chiefly be concerned with singly ionised species, though higher levels of ionisation are possible e.g. CO^{++} which results when CO loses 2e⁻. In summary, figure 1.4.

loses e⁻ to become

Figure 1.4   Formation of an ion.

1.4 MASSES OF ATOMS AND MOLECULES

The mass of an atom is the sum of the masses of its constituent protons, neutrons and electrons. That of a molecule is the sum of the masses of its atoms. The masses of atoms and molecules are minute, but a useful relative scale may be set up by expressing atomic and molecular masses in terms of an atomic mass unit (amu). 1 amu = 1.66×10^{-27} kg. With sufficient (whole number) accuracy for our purposes, relative atomic masses are then as shown in table 1.1.

Table 1.1 Relative atomic masses.

H	He	C	N	O	Ne	A
1	4	12	14	16	20	40

Relative molecular masses may be similarly expressed. Thus table 1.2.

Table 1.2 Relative molecular masses.

H_2	H_2O	N_2	O_2	CO	CH_4	NH_3	CO_2
2	18	28	32	28	16	17	44

In large amounts gases (and indeed matter in all its forms) are usefully specified in terms of a standard large amount, the *mole*. This is the amount of any material which contains the large number 6.02×10^{23} of its constituent atoms or molecules. 6.02×10^{23} is Avogadro's number, denoted N_A . The mass of one mole, the *molar mass*, is also known in older usage as the gram atomic weight or gram molecular weight, as appropriate. The molar mass of carbon is 12 g = 0.012 kg, of water 18 g = 0.018 kg. A mole of hydrogen gas weighs 2 g, a mole of oxygen gas 32 g, a mole of carbon dioxide 44 g etc. Refer to table 1.2. Avogadro's law asserts that, when measured at the same pressure, molar amounts of different gases occupy equal volumes.

With the knowledge that the density of liquid water is 1.0 g cm^{-3}, so that 18 g of water, a mole, occupies 18 cm^3 and contains 6.02×10^{23} molecules, we can assign the cubical volume occupied by a molecule of water to be $18/(6.02 \times 10^{23}) = 30 \times 10^{-24}$ cm^3 = 30 Å^3. If we imagine the water molecule as a sphere just fitting inside this cubical volume its diameter will be $(30)^{1/3} = 3.10$ Å, in accord with our earlier assertion about typical molecular sizes.

We shall, in later analysis, be concerned with the mass m of individual molecules. Its value can be deduced as follows. The molar mass M contains N_A molecules. Therefore

$$N_A \times m = M$$

So for hydrogen $m_{H2} = 2/(6.02 \times 10^{23})$ g $= 3.3 \times 10^{-27}$ kg
and for water $m_{H2O} = 18/(6.02 \times 10^{23})$ g $= 3.0 \times 10^{-26}$ kg

Henceforth, to avoid repeated usage of the phrase 'atoms (or molecules)' we will refer to the basic particles of a substance as 'molecules'.

1.5 BONDING, ENERGY AND TEMPERATURE. SOLIDS, LIQUIDS AND GASES

The primary effect at a molecular level of raising the temperature of a substance is to increase the energy of motion of its constituent molecules. The vigour of this motion opposes bonding.

In the solid state, molecules are held together in an ordered structure by bonding forces between the molecules which are similar in nature to those which operate inside the molecule between its atoms. They are however weaker. Thus, for example, water, H_2O, in the form of ice melts to become liquid water and not hydrogen and oxygen. Figure 1.5(a) depicts a simple model of an ordered solid structure. At a given temperature the molecules of the solid vibrate about the positions shown in a random and chaotic way. The energy of this microscopic motion is described as the thermal energy of the solid. The bonds restrain the motion and tend to be disrupted by it.

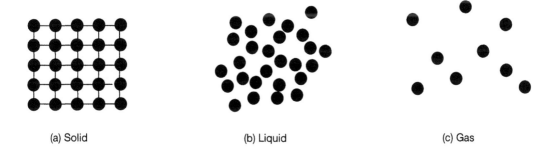

(a) Solid   (b) Liquid   (c) Gas

Figure 1.5   Structure of the solid (a), liquid (b) and gaseous (c) states.

As the temperature of a solid is raised, and the energy of vibration is increased, the bonding progressively weakens until, at the melting temperature of the solid, directional ordering is lost and the resulting liquid, though still consisting of molecules in close proximity, has a disordered structure, figure 1.5(b). Individual molecules diffuse and move easily around their neighbours and there is no rigidity in the whole assembly. Liquids flow.

With further increase of temperature, and further increase of energy of motion, the molecules of the liquid, already freed from the ordered arrangement, break their bonds completely and fly apart to become free particles occupying a much larger space, in total, than they did as liquid or solid. This is the gaseous state, figure 1.5(c). The increased energy of the molecules of the gas is now energy of free translational motion rather than the energy of restricted motion in the solid and liquid states. To emphasise the character of the gaseous state it may be recalled that a 1 cm cube of ice melts to a puddle of water of volume 1 cm^3, but this would vaporise to become 1700 cm^3 of steam. The book by Guinier (1984) is a good introduction to the molecular description of the solid, liquid and gaseous states.

We are now in a position to assemble our working model of a gas - the 'kinetic description' of a gas.

1.6 KINETIC DESCRIPTION OF A GAS

Our model of a gas is of free independent molecules of mass m moving in random directions, colliding frequently and exchanging energy in the collisions so that their individual directions and speeds are continually changing.

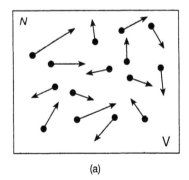
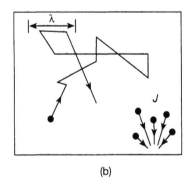

(a)   (b)

Figure 1.6 (a) Molecular motion; (b) mean free path and impingement.

Figure 1.6(a) depicts an enclosure of volume V containing N molecules. The number density n of the molecules in the gas is $n = N/V$. The figure may be regarded as a 'snapshot' of the state of affairs in the gas at a given instant. The different lengths of the arrows associated with each molecule are intended to convey that there is a range of molecular speeds and directions. Time lapse photography, if it were possible in these circumstances (which it is not), would show blurs of different lengths corresponding to this range of speeds and directions.

In figure 1.6(b) top left we imagine that we follow the progress of a particular molecule in time. It suffers collisions and random changes of direction and travels some average distance λ between collisions. This important quantity is called the *mean free path*. In the figure it would be a distance of about the length shown at the top left. At the bottom right of figure 1.6(b) the cluster of arrows represents molecules which, with a range of speeds and coming from all directions within the gas, impinge on the vessel wall. The total number of molecular impingements on unit area of the vessel wall in one second is an extremely important quantity, and is given the symbol J. It is most frequently referred to as the *impingement rate* but is also known as the particle *flux*. This latter usage is in a sense better, for the important reason that J represents, equally, the number of molecules which pass from one side to the other through imaginary unit area within the gas in one second. An equal flux, of course, passes in the opposite direction. J is conveniently measured as a number per cm^2 per second, and we shall describe it as an impingement rate or a flux according to what is appropriate in particular circumstances.

To recapitulate, our picture of a gas is of N molecules occupying volume V with number density $n = N/V$. The molecules of mass m move randomly with speeds v which reflect their energy of motion at temperature T. They collide with each other and with the walls of the containing vessel. The quantities λ, mean free path, and J, impingement rate, characterise the motion.

When this model of gas behaviour is analysed, a number of important relationships emerge between the quantities discussed above. They have been tested by experiment, and proven correct, so that the kinetic description of a gas is a good one. It was developed in the nineteenth century and is particularly associated with Clausius, Maxwell and Boltzmann. In the analysis the molecules are assumed to behave as hard elastic spheres of negligible size in comparison with the distances between them and the results which emerge are those for a 'perfect gas'. Real gases at low pressure, and therefore under 'vacuum' conditions, approximate very closely to this idealised state.

An excellent modern account of kinetic theory may be found in the book by Pendlebury (1985), one of the *Student Monographs in Physics* series. The physical chemistry texts by Atkins (1995, 1998) may also be consulted with advantage.

1.7   RESULTS FROM KINETIC THEORY

The principal results which emerge from the theory concern the distribution of molecular speeds and the dependence of pressure p, mean free path λ and impingement rate J on the number density n.

1.7.1   Molecular speeds (the Maxwell distribution)

Figure 1.7 shows the distribution of speeds at two temperatures. The distribution curve at a given temperature quantifies the picture presented in figure 1.6(a). It represents, at a given instant, the number of molecules which have a particular speed. The peak of the curve gives the most probable value of the speed and most molecules have speeds of this order. But some, a small proportion, are moving relatively slowly, and a similarly small proportion relatively fast. The average speed is shown as $\bar{v}$. It must be stressed that while this distribution of speeds is true at each and every instant in time, the identity of molecules which move with a particular speed is continually changing as collisions occur. A particular molecule, due to the collisions it suffers, will sometimes gain, sometimes lose energy so that, on average, its speed is $\bar{v}$.

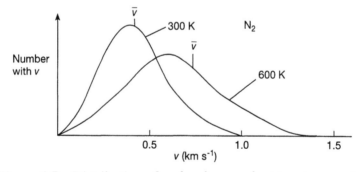

Figure 1.7   Distribution of molecular speeds at two temperatures.

1.7.2   Average kinetic energy of a molecule

$$\frac{1}{2}m\overline{v^2} = \frac{3}{2}kT \qquad (1.1)$$

This relates the average kinetic energy of a molecule to temperature. $\overline{v^2}$ is the mean square velocity, k is Boltzmann's constant $= 1.38 \times 10^{-23}$ J K^{-1} and T is the temperature in degrees K = degrees C + 273. The value of kT at room temperature taken as 22 °C is 4.1 × 10^{-21} Joule = 0.025 eV, i.e. 1/40 eV approximately. One eV unit of energy (=1 electron volt) = 1.6 × 10^{-19} Joule.

1.7.3   Average speed, $\bar{v}$

$$\bar{v} = \sqrt{\frac{8kT}{\pi m}} = \sqrt{\frac{8RT}{\pi M}} \qquad (1.2)$$

R is the universal gas constant, of value 8.314 Joule mole^{-1} K^{-1}. Inserting numerical values leads to

$$\bar{v} = 145 \sqrt{\frac{T}{M}} \text{ m s}^{-1} \qquad (1.3)$$

where M is the relative mass number as given in tables 1.1 and 1.2.

Note that (1) the higher the temperature T, the greater the average speed and (2) for different gases the average speed will be greatest for the gas whose molecules have the least mass (smallest M). Some typical values of $\bar{v}$ at room temperature, 22 °C (=295 K), are for hydrogen 1760 m s^{-1}, for helium 1245 m s^{-1}, for nitrogen molecules 470 m s^{-1} and for water vapour 587 m s^{-1}.

1.7.4   Pressure p exerted by a gas

$$p = nkT \qquad (1.4)$$

This result reflects the fact that each impact of a molecule with the wall exerts a force on the wall. The total pressure exerted depends on the number of impacts and the momentum (=mass × velocity) with which molecules hit the wall, hence the dependence on n and T. The number density n is the quantity of prime importance in vacuum practice as we shall see later.

Note that, at a given temperature, the pressure exerted by a gas depends only on the number density n of its molecules and not on their chemical identity. Thus, for example, helium and nitrogen gas exert the same pressure at a given temperature if their n values are equal, even though their molecules have different mass. The reason is that the lighter helium molecules although they travel faster on average and have a higher impingement rate on the wall of their container (see 1.7.6 below) have smaller momentum.

The value of n for a gas at room temperature and atmospheric pressure, taken as 295 K and 10^5 Pa respectively, is $n = p/kT = 10^5/4.1 \times 10^{-21} = 2.5 \times 10^{25}$ m^{-3} approximately (or equivalently, 2.5×10^{19} cm^{-3}).

1.7.5   Mean free path, λ

$$\lambda = \frac{1}{\sqrt{2}\,\pi d^2 n}$$ (1.5)

where d is the diameter of the molecules and the term πd^2 represents a collision cross section. Note that, as one might expect, λ increases as the number density of molecules (and hence the pressure) decreases. Molecular diameters of common molecules are listed by Kaye and Laby (1995). The diameter of a nitrogen molecule is 3.7 Å. For nitrogen gas at room temperature and pressure λ is a distance of 6.6 $\times$ 10^{-8} m or about 200 molecular diameters.

It is instructive to obtain an estimate of the average molecular separation in a gas at atmospheric pressure by imagining that, at an instant, the 2.5×10^{19} molecules which occupy 1 cm^3 are distributed roughly uniformly throughout the volume so that each is inside a cubic cell of volume $1/(2.5 \times 10^{19}) = 4 \times 10^{-20}$ cm^3. The length of the edge of this cell, namely $(4 \times 10^{-20})^{1/3} = (40 \times 10^{-21})^{1/3} = 3.4 \times 10^{-7}$ cm $= 34$ Å is a measure, albeit approximate, of average molecular separation. It is of the order of ten molecular diameters and, naturally, less than the mean free path under the same conditions.

1.7.6   Impingement rate/flux, J per unit area

$$J = \frac{n\bar{v}}{4}$$ (1.6)

Substituting for n from (1.4) and $\bar{v}$ from (1.2) gives

$$J = \frac{p}{\sqrt{2\pi mkT}}$$ (1.7)

This is frequently a more useful form of equation (1.6) when data for p and T are available in vapour pressure tables. Using $M = mN_A$ and $kN_A = R$ (the gas constant) leads to a further form:

$$J = \frac{pN_A}{\sqrt{2\pi MRT}}$$ (1.8)

1.7.7 Collision frequency

The collision which terminates the free path of a particular molecule, sending both it and the one it struck off in new directions for further collisions, is an unceasing and frequent event. We can estimate how frequent by dividing the average distance travelled in one second, $\bar{v}$, by the mean free path λ. Thus for nitrogen at room temperature and atmospheric pressure the collision frequency will be $470/6.6 \times 10^{-8}$ = 7.1×10^{9} per second, or, put another way, at this pressure a molecule spends an average time of order 10^{-10} second in travelling its free path between one collision and the next. At low pressures collisions are less frequent corresponding to longer free paths. However it is apparent that the pressure must be reduced by about ten orders of magnitude for the interval between collisions to be as long as one second, (and the volume occupied by the vacuum would then have to be rather large for collisions to take place!).

In summary of this section, the kinetic theory affords vivid insights, numerically based, into the behaviour of gases. We shall make considerable use of these results, especially those concerning n, λ and J, in section 2.2.

1.8 VAPOUR PRESSURE, EVAPORATION AND CONDENSATION

As previously discussed the molecules of a solid or liquid possess thermal energy which keeps them in a continuous state of agitation under the restraint of their binding forces. At the surface of a solid or a liquid, the molecules are less restrained than in the bulk because of their asymmetrical surface environment. At a rate which depends on temperature and the state of binding at the surface, molecules may break their bonds, leave the surface and enter the gas phase. There is therefore a continuous evaporation flux from the surface, though the magnitude of the effect varies over an enormous range, depending on the material and its temperature. Liquids such as ether for example evaporate rapidly as is well known. Water evaporates relatively slowly. Solids release their molecules in the same way, though at a rate which is many orders of magnitude smaller at room temperature. For evaporation rates from solids to be appreciable their temperature usually has to be quite high, a fact that is exploited in evaporation sources for thin film deposition.

A number of useful results may be deduced by considering the equilibrium of a liquid (or solid) with its vapour in a closed space maintained at constant temperature, as shown schematically in figure 1.8. When the arrangement is first set up, there is just liquid in the bowl and no vapour in the space above it. But as molecules escape from the liquid surface they start to build up a vapour. Some of these molecules in the vapour return to the liquid - at their prevailing density n as a vapour at any instant they will have a certain impingement rate back onto the liquid surface from which they came. So there is a constant evaporation flux J_E from the liquid, and a condensation flux J_C which builds up from nothing.

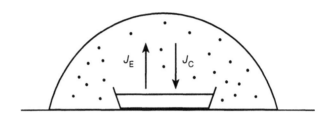

Figure 1.8   Equilibrium between a liquid and its vapour.

When the condensation flux has increased to a maximum   value J_C equal in magnitude to J_E, a state of equilibrium exists and the container is occupied by saturated vapour with a constant molecular number density n. In this state the vapour exerts its saturation vapour pressure (SVP), often simply referred to as its vapour pressure. Although on a laboratory scale nothing seems to be happening once equilibrium is reached, at a molecular level two vigorous processes are in balance.

By equation (1.7) with $J = J_C$ the condensation flux will be

$$J_C = \frac{p_e}{\sqrt{2\pi mkT}} \qquad (1.9)$$

where p_e is the saturation vapour pressure at temperature T. Therefore the evaporation flux J_E is also given by this expression. In non-equilibrium evaporation conditions, for example where the vapour released from a surface is being pumped away, equation (1.9) gives the maximum possible evaporation flux from the surface at temperature T. In general terms, the vapour pressure of a substance is a measure of its 'urge' to evaporate. The equation of this maximum evaporation rate as a mass flow rate for a material of molar mass M at temperature T is derived in appendix A.

The vapour pressure of a substance increases very rapidly with temperature because the release of molecules from a solid or liquid surface is a thermally activated process whose probability rises dramatically with increase of temperature. This thermal activation is discussed in detail in section 2.4.1 in connection with the process of desorption. The mechanism of release of the molecule is the same in both cases. Figure 1.9(a) shows on a linear scale of pressure the vapour pressure of water as a function of temperature - note the rapid increase of p with T. Figure 1.9(b) illustrates the way in which vapour pressure data are usually plotted using a logarithmic pressure scale. It is useful to remember that the vapour pressure of water at 22 °C is 23 mbar.

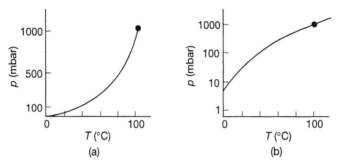

Figure 1.9   Vapour pressure of water as a function of temperature.

In the important non-equilibrium application of *vacuum drying*, in which water vapour is removed from wet pharmaceutical pastes by pumping, it is important to maintain the temperature of the surfaces involved at the designated value by supplying heat energy to provide the latent heat of vaporisation. Otherwise cooling occurs due to pumping, the vapour pressure is reduced, and the efficiency of removal falls. In other applications the cooling may be desirable. For example in low temperature physics, pumping hard over thermally well isolated liquid helium is used to achieve temperatures below 4 K.

In the non-equilibrium process of condensation of a gas or vapour on to a colder surface, such as arises in vapour traps and condensation pumping, equation (1.9) also gives the maximum possible condensation rate per unit area, assuming that the vapour pressure of the condensed species at the trap temperature is negligibly small, which is frequently the case.

1.9 GASES AND VAPOURS

The distinction as to whether matter in gaseous form is described as a gas or a vapour depends on its temperature, and is purely descriptive. If the material is above its critical temperature T_C it is described as a gas; below it as a vapour. The molecular behaviour - of free, randomly moving and colliding molecules - is the same in both cases. The essential point is that a vapour can be liquefied by compression, but a gas cannot. Below the temperature T_C, compression causes condensation of liquid droplets within the gaseous phase; above it compression causes a gas to become increasingly dense, but a liquid phase boundary does not appear. Consider the state of affairs shown schematically in figure 1.10, in which a vapour is contained in a cylinder closed by a piston and maintained at a temperature less than T_C. In each of the states of equilibrium depicted, the total downward force F exerted by the piston is equal to the upward force (= pressure × area) exerted on the piston by the vapour. The pressure arises, as discussed, from molecular impacts on the piston.

In figure 1.10(a) the vapour is unsaturated so that the pressure exerted is less than the saturation vapour pressure, p_e. As the volume is reduced to another state of equilibrium (with the heat energy arising from the work of compression being removed so that the temperature stays the same) an increased force is needed to maintain equilibrium against the increased pressure exerted by the vapour. The same number of molecules occupy a smaller volume so that their number density n is increased with a corresponding increase of pressure $p = nkT$. For the volume at which the vapour just becomes saturated (figure 1.10(b)) a force F_0 balances that due to p_e, and further reduction of volume occurs at constant force F_0 and vapour pressure p_e with condensation of some of the vapour to become liquid (figure 1.10(c)).

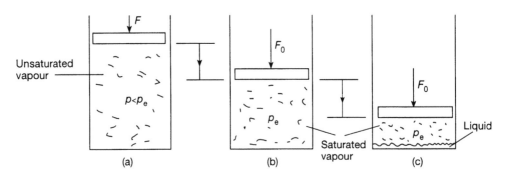

Figure 1.10 Equilibrium states of a vapour.

By contrast, at a temperature greater than T_C, compression of gas cannot produce condensation. With increase of pressure the volume decreases in accordance with Boyle's law though at high pressures where the volume of the molecules themselves becomes a significant fraction of the total volume they occupy, there are increasing deviations from ideal gas behaviour. Vapours also obey Boyle's law to a good approximation, provided they do not reach saturation conditions when, as depicted in figures 1.10(b) and (c), volume diminishes at constant pressure.

The critical temperature and associated conditions of pressure and molar volume are discussed by Guinier (1984) and a listing of these quantities for common substances is given by Kaye and Laby (1995). For example, the critical temperatures of nitrogen and water are 126 and 647 K respectively, reflecting our common perception of these materials at room temperature (~300 K) in gaseous form as gas and vapour respectively.

1.10 MACROSCOPIC GAS LAWS

The behaviour of real gases approximates more or less closely, depending on conditions, to the ideal gas laws associated with the names of Boyle and Charles and summarised in the statement that for a given mass of gas

$$pV / T = \text{constant} \tag{1.10}$$

This equation contains Boyle's familiar law 'pV = constant for a given mass of gas at fixed temperature'. The connection with the result $p = nkT$ already discussed arises by substituting $n = N/V$ to give

$$pV = NkT \tag{1.11}$$

For 1 mole $N = N_A$ and $N_A \times k = R$, the universal gas constant, of value 8.314 Joule mole^{-1} K^{-1}. Whence

$$pV = RT \tag{1.12}$$

This is the equation of state for 1 mole of ideal gas. Under conditions of standard temperature and pressure (STP), namely 0 °C (273K) and 1013 mbar (1.013×10^5 Pa), 1 mole of ideal gas occupies a volume 22.4 litres. This is often used with equation (1.10) as a starting point for simple calculations. Thus for example the volume of a mole at 22°C and standard pressure is $22.4 \times (295/273) = 24.2$ litres.

An arbitrary quantity of gas of molar weight M may be specified by its mass W or as a number of moles $n_M = W/M$. For example 64 g of oxygen is 2 moles. By extension of equation (1.12) for one mole it is clear that for n_M moles,

$$pV = n_M RT \tag{1.13}$$

The *standard litre* is the quantity of gas which occupies a volume 1 litre at STP, so it will be a mass equal to the molar weight divided by 22.4. A standard litre of oxygen therefore has mass 32/22.4 = 1.43 g, a standard litre of argon 40/22.4 = 1.78 g. A flow rate expressed as standard litres per minute (slm) of a specified gas is a <u>mass</u> flow rate, and once stated in these terms, is independent of the actual pressure and temperature conditions of the flow. Thus a flow of 1 slm at 506.5 mbar and 0 °C is a volumetric flow rate of 2 litres per minute, and would be a mass flow rate of 1.43 g per minute for oxygen. (See appendix H.)

1.11 GAS MIXTURES AND PARTIAL PRESSURES

In a mixture of gases A, B and C the total pressure p is the sum of the partial pressures p_A, p_B and p_C. A partial pressure is defined as the pressure that a gas would exert if it occupied the volume of the mixture on its own. Thus

$$p = p_A + p_B + p_C \qquad (1.14)$$

To see how the concept of a partial pressure arises consider the following situation. Suppose that, in order to make a mixture of gases A, B and C at a given pressure, atmospheric pressure P_0 say for illustrative purposes, one procures volumes V_A, V_B and V_C of the separate constituents, all at this pressure, and an evacuated container of volume $V = V_A + V_B + V_C$, of just the volume needed to contain the mixture. If gas A were introduced into the container on its own then by Boyle's law, it would exert a pressure p_A given by

$$p_A \times V = P_0 \times V_A$$

so that

$$p_A = (V_A/V)P_0$$

Similar results would apply for gases B and C considered separately:

$$p_B = (V_B/V)P_0 \qquad\qquad p_C = (V_C/V)P_0$$

When all three gases at P_0 are introduced into the volume V and mixing has occurred the total volume is unchanged and they exert partial pressures in accordance with equation (1.14) so that, in this example, the sum of partial pressures is P_0.

If $n_{M,A}$, $n_{M,B}$ and $n_{M,C}$ are the number of moles of each constituent in the volume V then from equation (1.13) $p_A = n_{M,A}(RT/V)$ etc, and so we can write

$$p = (n_{M,A} + n_{M,B} + n_{M,C})RT/V \qquad (1.15)$$

As an example consider 100 g of a helium/oxygen mixture made up of 4 g of He and 96 g of O_2. There is one mole of helium and three moles of oxygen in the mixture, and so one atom of helium for every three of the heavier oxygen molecules. The partial pressures of the constituents will be proportional to their molar concentrations and so for any conditions of the gaseous mixture the partial pressure of oxygen will be three times that of helium. For example if the volume and temperature are such that the total pressure is 100 mbar, the partial pressures of oxygen and helium will be 75 and 25 mbar respectively.

In molecular terms, each partial pressure can be associated with a molecular number density so that $p_A = n_A kT$ etc with the total n given by

$$n = n_A + n_B + n_C$$

Atmospheric air is a mixture of gases and water vapour. The amount of the latter varies considerably with atmospheric conditions. At a given temperature the relative humidity (RH) expresses the amount present as a fraction of the maximum amount possible, corresponding to saturation. Environmental temperature and humidity are both important in determining bodily comfort. Comfortable humidity at 20 °C is about 60%. For RH values much greater than this, the body's ability to cool by perspiring is diminished and the air feels 'muggy'. For RH values below about 35% the air feels raw and causes throat irritation.

The main constituents of dry air are given in table 1.3 (data taken from Kaye and Laby 1995).

Table 1.3 Main constituents of dry atmospheric air.

Gas	Volume (%)	Partial pressure (mbar)
Nitrogen	78.08	791.1
Oxygen	20.95	212.2
Argon	0.93	9.4
Carbon dioxide	0.03	0.3

Since the vapour pressure of water at 22 °C is 23 mbar it is clear that with typical RH values of about 50% water vapour is present in the atmosphere in significant amounts, exerting a partial pressure of order 1% of the total pressure.

1.12 CONTINUUM AND MOLECULAR STATES OF GAS, KNUDSEN NUMBER Kn = λ/D

From the kinetic description of a gas we have learned that for a gas at atmospheric pressure the mean free path λ is very small, of order 10^{-7} m for typical gases. At low pressures of course when a gas has a smaller molecular number density n the mean free path is larger. For example, at a pressure of 10^{-5} mbar the mean free path (which is to be discussed in much more detail in chapter 2) is about 7 m. This implies that to observe a gas in the state described in section 1.6 - of chaotically moving molecules undergoing incessant mutual collisions - the volume of the container enclosing it would have to be rather large, in this instance with a diameter of at least 100 m say, if it were spherical.

When gases do exist at number densities n which imply λ values greater than the size of their containers, a number of new and interesting phenomena arise which are of central importance for vacuum practice, and which will be dealt with in due course. Under these conditions gas molecules collide with the surfaces of their containing boundaries more frequently than with each other. Two numbers should therefore be specified to characterise the state of a gas in a container: one, the mean free path λ, a characteristic property of the gas determined by its n value, the other a dimension D characteristic of its container. In the case of a cubical box D would be the length of an edge; for a long pipe D would be its diameter. Both λ and D have the units of length and so their ratio is dimensionless. It is known as the *Knudsen number,* symbol Kn, and is defined as Kn $= \lambda / D$.

When, as at atmospheric pressure, λ is very much smaller than typical container dimensions, the kinetic picture of gas behaviour is that presented in figure 1.6(a). Molecule - molecule collisions dominate gas behaviour and it behaves as a *fluid*. λ is much less than D so that Kn $\ll$ 1. Air in a car tyre is an example, though examples abound since this is the state of the atmosphere in which we live and breath. Only under these familiar fluidic conditions can pressure waves and therefore sounds can be transmitted. Gases for which Kn $\ll$ 1 are said to be in a *continuum* or *viscous* state.

By contrast, for a TV display tube of size $D \sim 0.5$ m in which there is a vacuum of typically 10^{-5} mbar, $\lambda \gg D$, Kn $\gg$1 and molecule - surface collisions dominate gas behaviour, which is <u>not</u> fluid-like. Gas in these conditions is referred to as being in a *molecular* state.

The Knudsen number Kn combines λ and D in a way that automatically incorporates the scale of the phenomenon involved. Thus gas with a λ value of 7 m in a small laboratory vacuum system would be in a molecular state. But to an atmospheric scientist concerned with volumes of size 1 km^3 in the upper atmosphere, it would be in a continuum state.

The ratio of the number of molecule - surface collisions on the interior bounding surface of a volume to the number of molecule - molecule collisions within it can be easily worked out for simple shapes, as shown below. Consider a cubic container of side L containing gas with number density n and associated mean free path λ. The number of collisions per second on the interior surface is $6L^2 \times (n\bar{v}/4) = 3n\bar{v}L^2/2$ using equation (1.6). The number within the volume is $L^3 \times (n\bar{v}/2\lambda)$, which follows from the discussion in section 1.7.7. Therefore dividing the former by the latter, the ratio of number of surface to the number of volume collisions is $3\lambda/L$, which is three times the Knudsen number, Kn. The ratio has a similar value for other shapes such as spheres and equi-axed cylinders. This is an important application of the Knudsen number because it tells us numerically, for a gas at a given pressure in a given container, whether molecular behaviour is dominated by inter-molecular collisions or by their collisions with the atoms or molecules of their confining wall. Thus, in contrast to the state of affairs at atmospheric pressure, for a cubic volume of side 0.5 m containing gas at 10^{-5} mbar ($\lambda = 7$ m), surface collisions are more frequent than inter-molecular collisions by a factor of $3 \times 7/0.5 = 42$. This result will be used again in chapter 2.

1.13 HEAT CONDUCTION IN A GAS

Conduction, convection and radiation are the processes by which heat energy may be transferred through a gas from a higher to a lower temperature. Thermal radiation such as one experiences when being warmed by an electric fire bar is in fact almost indifferent to the intervening gas and so the laws which control it, as discussed for example by Zemansky and Dittman (1997), apply equally in the context of vacuum practice as they do in more familiar situations. Convection is the process by which gas adjacent to a hot surface is warmed by conduction, expands, becomes buoyant with respect to its surroundings, and therefore rises carrying the heat energy acquired with it. This is called *free* convection. *Forced* convection occurs when a directed draught of gas achieves heat extraction in a similar way.

Thermal conduction in a gas may be understood in terms of the molecular-kinetic description presented earlier. The molecules of gas in equilibrium have a kinetic energy determined by the temperature T as expressed in equation (1.1). The higher the temperature the higher the average speed as equation (1.2) shows directly. The molecules of a gas encountering a surface which is hotter than they are depart from it on average faster than they arrived, having gained extra energy prior to departure from the vibrational energy of atoms in the (hotter) solid surface. But as they travel back into the gas with increased average speed they suffer collisions with other molecules after an average distance λ. In colliding with other molecules they lose some of this acquired energy, passing it on and transporting it away from the hot surface through the gas. Based on this mechanism a simple model of the thermal conduction process may be constructed (see, for example Walton (1983)).

The thermal conductivity κ of a material is defined as the ratio of the heat flow per unit area through it in a given direction to the rate of temperature change with distance in that direction, the temperature gradient, so that the greater the κ value the greater the heat flow for a given value of the temperature gradient. The kinetic model shows that $\kappa = 0.5 \lambda nm \, \bar{v} \, c$ where c is the specific heat per kg at constant volume. An important consequence of this expression, verified experimentally, is that κ does not depend on pressure. This is because the dependence of λ on $1/n$ (equation (1.5)) removes n, and hence p, from the formula for κ. Basically the effect of an increase of n which creates a greater number of energy carriers is exactly compensated by a corresponding shortening of the mean free path, so that the extra energy 'launched' travels a smaller distance.

This picture in which the conductive process depends on molecular collisions clearly relates to gas in a continuum state of small Knudsen number. For more rarefied gas as Kn increases and molecular conditions are approached κ ceases to be independent of pressure and decreases as pressure falls. This is exploited in gauges to be described in chapter 4.

1.14 VISCOSITY OF GASES

Imagine a large horizontal plate and gas, say air, above it. Suppose we arrange for the plate to move at a constant speed to the right in a horizontal direction. The gas molecules closest to the plate which bombard it and rebound upwards will tend to be moved along with it, having acquired an extra speed component to the right during the encounter. However as they leave the plate in a generally upward direction they will, after a distance of about λ, collide with other molecules which do not have this extra imposed motion towards the right, and so as a result lose some of their acquired motion. In doing so the struck molecules will gain some, tending themselves to be moved to the right, and so on. The process is closely related to that described in the previous section, but what is being transferred away from the surface in this case is *momentum* parallel to it, imposed by the motion of the plate, rather than energy.

If the motion of the plate is sustained adjacent layers of the gas acquire a motion in the same direction, which decreases progressively with distance from the surface of the plate, since the generally haphazard motion of the molecules opposes the orderly motion being imposed. Normally the velocities being imposed are very much smaller than molecular velocities. When the motion of the plate ceases, the relative motion in the gas does also. The property of gases which cause them to exhibit this internal resistance to ordered layer-over-layer type motion is called viscosity. It slows down the movement of objects through gas, and that of gas itself when pressure differences are applied to make it flow, through pipes for example. In a similar way to heat flow, a coefficient of viscosity η describes how the relative motion and the shearing force which sustains it are related. In terms of the molecular-kinetic model, η is related to other molecular quantities already defined by the formula $\eta = 0.5\lambda nm \ \bar{v}$. Here again, one may deduce that in the continuum (fluid) state, gas viscosity will not depend on pressure, an at first surprising result which had some historical importance (Walton 1983).

1.15   GAS - SURFACE SCATTERING AND MOLECULAR DRAG

As will be discussed more fully in section 2.6, when gas molecules impinge on a stationary surface they do not usually rebound elastically from it as would a billiard ball from the table edge, with angle of reflection equal to that of incidence, as depicted in figure 1.11(a). Rather, as shown in figure 1.11(b), they leave in a random direction determined by the complex forces which operate between the gas molecule and the atoms of the surface. The molecule is as likely to take off again in a 'forward' direction FWD as a 'backward' one BWD.

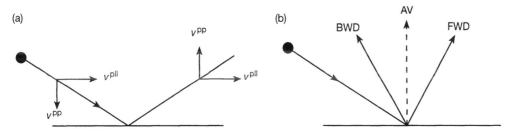

Figure 1.11   (a) Specular and (b) normal rebound of a molecule from a surface.

On average therefore the direction of departure is perpendicular to the surface, denoted AV in the figure. (The argument is made in two dimensions, but is true also in three.) Figure 1.11(a) is nevertheless useful in demonstrating how the velocity, and hence momentum of the incoming molecule may be resolved into components perpendicular and parallel to the surface v^{pp} and v^{pll} respectively. For a collision of the billiard ball type with equal angles of incidence and reflection - a so-called 'specular' reflection - v^{pp} is reversed while v^{pll} is unaltered.

In the typical case of figure 1.11(b), the molecules arriving in this direction bring momentum mv^{pll} to the surface but <u>on average</u> have no parallel momentum when they leave because they depart in the perpendicular direction AV. Overall therefore the parallel momentum mv^{pll} of an incoming molecule is destroyed and during the brief time in which it is brought to rest a force is exerted on the surface in the direction of the incoming motion. Since molecules arrive from all directions there is no net force in a sideways direction on the surface. (It is of course the reversals of the perpendicular components of momentum mv^{pp} occurring at a great rate per unit area of the surface which create the pressure on it.)

Consider now the situation in which the surface on which molecules impinge is not stationary. When the gas pressure is low so that Kn values approach or exceed unity, and the surfaces move at speeds of order molecular speeds, appreciable molecular drag effects are observed. Their origin may be understood by considering the hypothetical device shown in figure 1.12, which we imagine as part of a 'thought experiment'. A continuous belt is driven in vacuum at constant speed U comparable with molecular speeds, guided on rollers which are also supplied with power to maintain the motion. The mechanical arrangement resembles that of a belt sander. It is the molecules in the upper region between the top surface of the belt and the stationary plate which are of interest. We suppose that the belt - plate separation is less than the mean free path so that molecules leaving the belt cross to the plate without being scattered by other molecules, i.e. Kn > 1 and conditions are molecular.

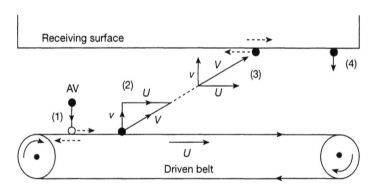

Figure 1.12   Device to illustrate molecular drag.

Consider the arrival of gas molecules on to the moving belt, event (1) in the figure. All arrival directions are equally probable and so the average direction of arrival AV is perpendicular to the belt as indicated. When it arrives at the belt the average molecule experiences a 'sideways' force to the right which accelerates it to the velocity U of the belt. There is an equal and opposite slowing force on the belt so that the motor has to do extra work to maintain the belt speed.

At a later time, event (2), this average molecule will depart from the belt surface, again on average in a perpendicular direction with respect to the belt with velocity v, but in addition with a sideways velocity U imposed by the belt. The total velocity will be V compounded of v and U, as shown. The belt thus drags molecules in the direction of its motion, and imposes an orderly component to the right on the otherwise haphazard molecular motion.

The average molecule arrives (3) at the upper surface bringing an imposed parallel momentum mU to it and, as previously discussed, in coming to rest gives up this momentum, creating a force which tends to move (drag) the plate to the right. This is the total parallel momentum change because again, on average, the direction of departure from the stationary plate will be perpendicular to it.

If we continue to follow the trajectory of our average molecule we see that the cycle of perpendicular arrival onto the belt, right-biased departure from it with arrival and momentum exchange at the upper surface, followed by downward perpendicular departure from the stationary plate, is continually repeated. The overall effect therefore is that of dragging the molecular flux incident on the belt towards the right. This effect is exploited in the 'molecular drag' stages of certain pumps to be discussed in chapter 3.

Thus three distinct aspects of molecular drag phenomena may be identified. First, the imposed bodily movement of gas discussed above. Second the tendency to drag a stationary object, in this example the upper plate, in the direction of the biased motion. Third, the slowing effect on fast moving surfaces due to the arrival of random molecular flux onto them, exemplified in this case by retarding forces on the belt. This effect is exploited in one type of pressure gauge to be discussed in chapter 4.

These effects are analysed with relative ease for gas in a molecular state, and two examples are given in appendix B. As might be expected their magnitude is directly proportional to the number density n, and therefore pressure. However at higher pressures, in transition to the continuum state, where as we have seen in section 1.14 the forces due to relative motion are described as viscous and become independent of pressure, matters become complicated. Their basis is well discussed by Walton (1983).

1.16 THERMAL TRANSPIRATION

We consider finally in this chapter a phenomenon that is relevant to the measurement of pressure in circumstances where there are temperature differences between parts of a vacuum system and gas is in a molecular state.

Consider as in figure 1.13 an arrangement of two vessels at different temperatures T_1 and T_2 containing gas at low pressures p_1 and p_2 and connected together through a very short pipe, so short that it can be considered as a simple aperture.

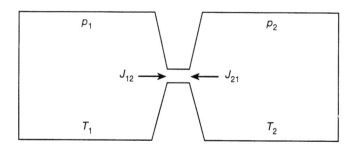

Figure 1.13 Illustrating conditions for thermal transpiration.

Suppose the pressures are such that the mean free path is greater than the aperture diameter so that the gas is not in a continuum state. The condition of equilibrium achieved under these conditions, determined by the fact that once a steady state is reached the molecular fluxes J_{12} and J_{21} in opposite directions must be equal, is found using equation (1.7) for J to be:

$$\frac{p_1}{\sqrt{T_1}} = \frac{p_2}{\sqrt{T_2}}$$

and so

$$\frac{p_1}{p_2} = \sqrt{\frac{T_1}{T_2}} \qquad (1.16)$$

The phenomenon described by this equation is called thermal transpiration. It is different from the equilibrium that would occur if the gas were in a continuum state, when equality of p_1 and p_2 would characterise the equilibrium. The classic text of Loeb (1961) discusses these matters in detail.

Returning to figure 1.13 and equation (1.16) imagine for example that region 1 is at liquid nitrogen temperature 77 K and region 2 at room temperature 293 K where a pressure gauge measures p_2. Then $p_1 = \sqrt{77/293}\, p_2 = 0.51 p_2$. Careful experimentation has verified equation (1.16) in equipment where the idealised situation of figure 1.13 is closely approximated. In many practical situations however analysis is complicated by the presence of tubulation and gradual rather than abrupt temperature changes between the region of interest and the location of a pressure gauge at room temperature. As a result ratios p_1/p_2 are larger than indicated by equation (1.16), i.e. the pressure in a cold region deduced from a room temperature gauge will be higher than (1.16) indicates. Poulter (1983) discusses practicalities in useful detail.

Summary

In this chapter we have described the behaviour of gases on a microscopic scale in terms of their molecular motion and shown how kinetic theory gives quantitative insights into this behaviour. Some of the more familiar macroscopic quantities have been discussed and related to the model. In addition, the concept of rarefied gas in a molecular state has been introduced, and some relevant properties of gas in this state discussed.

References

Atkins P W 1995 *The Elements of Physical Chemistry* 2nd edition (Oxford: Oxford University Press)

Atkins P W 1998 *Physical Chemistry* 6th edition (Oxford: Oxford University Press) chapters 1 and 24

(The first of these gives an excellent introductory account of the matters referred to. The second gives a more detailed, but equally accessible treatment.)

Guinier A 1984 *The Structure of Matter* (London: Arnold)

Kaye and Laby 1995 *Physical and Chemical Constants* 16th edition (London: Wiley)

Loeb L B 1961 *The Kinetic Theory of Gases* (New York: Dover)

Pendlebury J M 1985 *Kinetic Theory* (Bristol: Institute of Physics)

Poulter K F *et al* 1983 *Vacuum* **33**(6) 311

Walton A J 1983 *The Three Phases of Matter* 2nd edition (Oxford: Clarendon)

Zemansky M W and Dittman R H 1997 *Heat and Thermodynamics* (New York: McGraw-Hill)

2

Gases in Vacuum Systems

A Chambers

The purpose of this chapter is to set up a framework for the description of gas behaviour in vacuum systems, in preparation for later discussion of how vacua are created and characterised in practice. Of central importance is the description of gas flow and the specification of performance.

2.1 THE BASIC TASK, UNITS AND RANGES OF VACUA

The basic task of the vacuum engineer may normally be stated by reference to figure 2.1. It is to reduce the density of gas in a vessel to a value sufficient for its intended purpose. This will be achieved by connecting a pump to the vessel. The pump may expel the gas withdrawn to atmosphere safely, or store it in a condensed state.

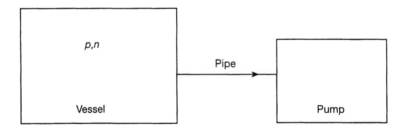

Figure 2.1 Schematic representation of a vacuum system.

Whatever the purpose in hand it will be the number density n of molecules in the vessel that determines the suitability of the vacuum for that purpose. The pressure p exerted by the gas is in most cases of no direct physical interest. It simply serves, via the relationship $p = nkT$, as a convenient measure of the quality of the vacuum.

Our measure of a vacuum then is the total pressure of residual gases in the vessel. It may be expressed in various units which are related by the following statement:

$$1 \text{ atmosphere } = 760 \text{ torr} = 1013 \text{ millibar} = 1.013 \times 10^5 \text{ pascal}$$

Although the pascal, Pa (= 1 N m^{-2}) is the SI unit, current practice in Europe continues to favour the millibar (mbar). This unit has the advantage that it is not very different from the obsolete, but still widely used unit, the torr, which is the pressure that will support a column of mercury 1 mm high. 1 mbar = 3/4 torr approximately. We will use the millibar following common practice. Since 1 mbar = 100 Pa, conversions to and from SI are easily made. The various ranges of vacua are defined by the scale shown in table 2.1. It should be noted that the scale of measurement is logarithmic, not linear, and spans many orders of magnitude.

Table 2.1 Defining the various ranges of vacuum.

p (mbar)	10^3		1		10^{-3}		10^{-7}		10^{-12}	
		-Low-		-Medium-		-High-		-Ultra-high-		Extreme high
						HV		UHV		XHV
p (Pa)	10^5		10^2		10^{-1}		10^{-5}		10^{-10}	

It is instructive to tabulate the values of number density n, mean free path λ and impingement rate J for representative pressure values on this scale. Table 2.2 shows data for nitrogen gas N_2, the dominant constituent of air, at 22 °C = 295 K.

Table 2.2 n, λ, and J at various p for N_2 at 295 K.

p (mbar)	n (m^{-3})	λ	J (cm^{-2} s^{-1})
10^3 = 1 atm	2.5×10^{25}	6.6×10^{-6} cm	2.9×10^{23}
1	2.5×10^{22}	6.6×10^{-3} cm	2.9×10^{20}
10^{-3}	2.5×10^{19}	<u>6.6 cm</u>	2.9×10^{17}
10^{-6} , HV	2.5×10^{16}	66 m	<u>2.9×10^{14}</u>
10^{-10} , UHV	2.5×10^{12}	660 km	2.9×10^{10}

Note that:

(1) The number density n varies over an enormous range. At 10^{-6} mbar, a conventional high vacuum, it is one thousand millionth of its value at atmospheric

pressure. But even at 10^{-10} mbar, UHV, the number density of 2.5 x 10^{12} per m^3 or equivalently 2.5 x 10^6 per cm^3, is very large.

(2) Mean free path naturally increases as pressure falls. At a pressure of 10^{-4} mbar where λ = 66 cm it has become comparable with the dimensions of typical containers.

(3) J is a very large number whatever the pressure. The value of J at 10^{-6} mbar, 2.9 $\times 10^{14}$ $cm^{-2}s^{-1}$, has particular significance for the following reason. As we shall later discuss in more detail, gas molecules may become attached to solid surfaces by the process of adsorption. The number of molecules that would form a complete molecular layer, known as a *monolayer*, on 1 cm^2 of surface is about 10^{15}. This may be deduced as follows. Imagine molecules of diameter 3 Å = 3×10^{-8} cm packed in a square array on 1 cm^2 of surface. In a line 1 cm long there would be $1/(3 \times 10^{-8})$ = 3.3×10^7 of them, and therefore $(3.3 \times 10^7)^2 = 1.09 \times 10^{15}$ molecules in a 1 cm square. Suppose, for example, that an area of 1 cm^2 of fresh surface is prepared in a vacuum of 10^{-6} mbar by cleaving a crystal to expose new surface, and assume that all the arriving molecules stick to it. Then, since J = 2.9×10^{14} from the table above, this surface would be completely covered with adsorbed gas in about 3 seconds at this pressure. The so-called *monolayer formation time* is therefore of order seconds at 10^{-6} mbar.

(4) In the discussion of section 1.12 it was shown that the ratio of the number of collisions per second of molecules on the bounding surface of the volume which contains them to the number of molecule - molecule collisions within it is $3\lambda/L$. From the λ values in table 2.2 one can deduce that for a vacuum in a cubic vessel of side 0.5 m, which is a typical size (though not shape), the rates would be comparable at 10^{-3} mbar, while at 10^{-6} and 10^{-10} mbar surface collisions would dominate by factors of order one thousand and ten million respectively. At these high vacua therefore surfaces exert a controlling influence on the state of the gas they contain. Within the body of the gas molecule - molecule collisions are relatively rare, and in most circumstances insignificant.

2.2 FORMULAS FOR IMPORTANT QUANTITIES
The entries in table 2.2 are obtained as follows using the results presented in section 1.7.

2.2.1 Number density, n
Since p = nkT with k = 1.38×10^{-23} J K^{-1}, then for T = 295 K,

$$n = p/kT = 2.5 \times 10^{20}p \ \text{m}^{-3}, \ p \text{ in Pa}$$

For p in mbar, since 1 mbar = 100 Pa,

$$n = 2.5 \times 10^{22}p \ \text{m}^{-3}, \ p \text{ in mbar} \tag{2.1}$$

2.2.2 Mean free path, λ

$$\lambda = \frac{1}{n\sqrt{2}\,\pi d^2}$$

Taking $d = 3.7 \times 10^{-10}$ m for the N_2 molecule and substituting for n from $p = nkT$

$$\lambda = (1/n\sqrt{2}\,\pi d^2)/(p/kT) = 6.6 \times 10^{-3}/p \quad \text{metre, } p \text{ in Pa}$$
$$= 6.6 \times 10^{-3}/p \quad \text{cm, } p \text{ in mbar}$$

With sufficient accuracy for approximate calculations we may take

$$\lambda = 7 \times 10^{-3}/p \quad \text{cm, } \quad p \text{ in mbar} \tag{2.2}$$

Thus at a pressure of 10^{-3} mbar, the mean free path is about 7 cm (~3 inch). Because its use arises very frequently, it is worthwhile committing this expression, or some version of it, to memory. Since the product λp is constant, remembering the value of λ at a particular pressure means that values at other pressures are easily computed.

The values of the mean free path for other gases may be computed if their molecular diameters are known. There is not a large variation in molecular size for the more common gases and so λ values at a given pressure are all comparable in magnitude. Thus at 10^{-3} mbar the values for helium and argon are 13.7 and 5.7 cm respectively compared with 6.6 cm for nitrogen. Roth (1990) discusses mean free path concepts in gas mixtures.

2.2.3 Impingement rate/flux, J

$$J = \frac{p}{\sqrt{2\pi mkT}}$$

$$= \frac{pN_A}{\sqrt{2\pi MRT}}$$

whence for nitrogen ($M = 0.028$ kg) at 295 K

$$J = 2.9 \times 10^{22} p \quad \text{m}^{-2}\text{s}^{-1}, p \text{ in Pa}$$
$$= 2.9 \times 10^{20} p \quad \text{cm}^{-2}\text{s}^{-1}, \ p \text{ in mbar}$$

Again, with sufficient accuracy for approximate calculations we may take

$$J = 3 \times 10^{20} p \quad \text{cm}^{-2}\text{s}^{-1}, \ p \text{ in mbar} \tag{2.3}$$

2.3 QUALITATIVE DESCRIPTION OF THE PUMPING PROCESS

Let us return to consider the basic task. To achieve the required vacuum is not simply a matter of removing a sufficient quantity of the air originally in the vessel. This indeed has to be removed but we then find that there are continuous sources which launch gas into the volume and which present the pump with a continuous gas load. (In fact in some low-vacuum applications gases are continuously injected into a vessel in order that reactions for example can take place, and gaseous products have to be removed.) Thus the vacuum achieved at steady state is the result of a dynamic balance between the gas load and the ability of the pump to remove gas from the volume. Figure 2.2 illustrates the situation.

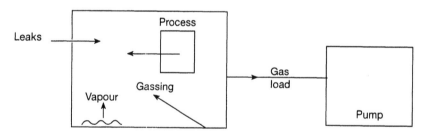

Figure 2.2 Sources of gas which present a load to the pump.

We may divide these sources into broad categories, and, to anticipate section 2.5.2, denote the loads they present by the symbol Q.

(1) **Leaks, Q_L.** These may be real leaks due to passageways through the vacuum wall from outside the vessel or, more subtly, virtual leaks due to gas being trapped in localities from which it can emerge only slowly into the vacuum surroundings. The deliberate injection of gas into a vessel which, as mentioned previously, is a feature of some processes, would also constitute a leak in the sense used here. See section 7.1 for more discussion of real and virtual leaks.

(2) **Vaporisation, Q_V.** Often inadvertently, but sometimes of necessity, materials which exert a significant vapour pressure are present in a vessel, contributing a gas load. Water vapour from imperfectly dried components is, in general, troublesome because it slows down the initial evacuation process. (There are, on the other hand,

some applications such as vacuum drying of wet pharmaceutical pastes where water vapour removal is the prime purpose of the vacuum process.) Careless handling of components which leaves organic material such as fingerprints on them should be strictly avoided in HV and UHV practice; not only do they evaporate slowly, they can become carbonised in electron beams and cause electrical breakdown problems on insulators.

(3) **Outgassing, Q_G.** This term describes the release of gas from the internal surface of the vacuum wall and the surfaces of components inside the vessel. It forms the principal source of gas in many systems and limits the degree of vacuum which can be achieved. Its origins and magnitude are discussed in section 2.4.

(4) **Process generated gas, Q_P.** Many processes carried out in vacuum cause the release of gas, often from materials which are heated. For example in vacuum degassing applications metals are heated to high temperature to rid them of dissolved gas.

(5) **Others.** Depending on the type of pump used in a given application, there may be a tendency for the vapour of lubricants or, in the case of a diffusion pump, vapour of the working fluid, to 'back-stream' into the vacuum chamber. Precautionary measures such as interposed traps may be used to reduce this, as will be described in chapter 3.

As will be explained later in more detail, pumps do not work perfectly and capture all molecules which enter them, so that some molecules from the vacuum chamber return to it and the 'gas load' shown entering the pump in figure 2.1 is a *net* load removed from the chamber.

In summary, gas from various sources is launched into the vacuum space. The effectiveness of the pump in removing it determines the vacuum achieved. At steady state the pressure of residual gases in the vessel reflects a dynamic balance between pump performance and gas load.

2.4 SURFACE PROCESSES AND OUTGASSING

This is a large subject and the purpose of this section is to present its essential features in a way which will enable it to be pursued further in the references cited. The articles by Elsey (1975) remain a valuable contribution to the literature. More recently Berman (1996) has given a comprehensive description of the role of water vapour in vacuum systems.

Outgassing is the release of gas from internal surfaces into the vacuum. It occurs by *desorption* of molecules from bound states on the solid surface. That a bond may be established between a gas molecule and a surface should not seem too surprising. An atom in the surface of a solid has fewer neighbouring atoms than an atom in the interior and therefore has some spare bonding capacity, so to speak, which may be used to establish a bond with a 'foreign' atom or molecule. A molecule thus attached to a surface is said to be *adsorbed*.

Molecules impinging directly on to the surface from the gas phase may be captured by the process of adsorption. Also, of equal significance for vacuum purposes, as we shall see, they arrive at the surface by <u>diffusion from the interior of the material of the vacuum wall</u>. Hydrogen, oxygen, nitrogen and carbon oxides are dissolved into the bulk of structural materials such as stainless steel in the course of their manufacture, and in surprisingly large amounts. A one centimetre cube of stainless steel may contain absorbed gas that would occupy a comparable volume at atmospheric pressure.

It is of first importance for our purposes to consider the process of desorption. The reverse process, adsorption, will be discussed briefly in due course.

2.4.1 Desorption and adsorption

Consider figure 2.3(a) which depicts an adsorbed molecule attached to a surface by a bond. In an energy - distance diagram, figure 2.3(b), the molecule is trapped in a potential energy well of depth q which represents the energy of binding at an equilibrium distance r_0 from the surface. The molecule would have to acquire energy of at least this amount q to break the bond and escape from the surface, i.e. to desorb.

Now the surface atoms and the molecule exist at finite temperature T and so, as discussed in section 1.5, have thermal energy which causes them to be in a state of continuous vibratory motion about their equilibrium positions. Individual motions are haphazard, and the adsorbed molecule is being continually and randomly 'buffeted' by contact with its neighbouring atoms, so that energy is shared back and forth between them, and the adsorbed molecule vibrates around its equilibrium position r_0. It will eventually happen by statistical chance that the motions of the atoms in the surface effectively assist each other and transfer enough energy to the adsorbed molecule, q or more, so that the bond breaks and the molecule desorbs. The probability of this happening depends on the magnitude of q and the temperature T via the 'Boltzmann factor', $\exp(-q/kT)$. The greater the bond strength q the smaller the probability of desorption at a given temperature; for a given q the probability will increase with temperature because of the increased thermal motion. A process of this sort, which depends on fluctuations of thermal origin, is described as being 'thermally activated'.

It is appropriate now to consider the magnitude of the energies q. They are frequently specified in kilojoules per mole. Using the fact that a mole contains N_A molecules, it is easily shown that <u>97 kJ   per mole is equivalent to 1 eV per molecule</u>.

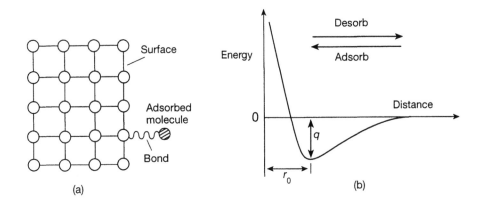

Figure 2.3 The adsorbed state.

Depending on the identities of the gas molecule and the surface, an adsorbed molecule may be either physically sorbed or chemisorbed. In the former case the bonding is weak, being due to fluctuating electric dipole forces of the van der Waals type, and the molecule's structure is substantially unaltered. The weakness of the bonding is reflected in the relatively small heats of adsorption H_A, which lie in the range 8 - 40 kJ per mole. For example it is about 12 kJ per mole for nitrogen on glass. H_A is the heat energy released on adsorption. In terms of the energy diagram of figure 2.3(b), the energy barrier to the molecule's desorption is $q = H_A/N_A$. In chemisorption, bonding is generally much stronger and involves the transfer or sharing of electrons in bonds which are of the same type as those which hold atoms together inside the molecule. Indeed the process is so named because it is a chemical reaction between the adsorbed molecule and those of the surface. In many cases the adsorbing molecule dissociates. The energy - distance diagram is then more complex than figure 2.3(b) and heats of adsorption H_A and energy barriers for desorption are higher, in the range 60 - 800 kJ per mole. For example for hydrogen on iron H_A is 134 kJ per mole.

Returning to consider the state of an adsorbed molecule recall that, once adsorbed, it is subject to perturbing effects which will cause it eventually to desorb. It follows that adsorbed molecules have limited lifetime on the surface. However remembering that kT at room temperature is 1/40 eV we see that this average thermal energy is significantly less than desorption energy barriers. That is why fluctuations sufficient to provide the release energy q are rare. The average lifetime of an adsorbed molecule is called the mean adsorption stay time, or sojourn time, τ. De Boer (1953) gives a very clear and informative discussion based on a simple model. According to this model $\tau = 10^{-13}\exp(q/kT) = 10^{-13}\exp(40q)$ seconds at room temperature, where 10^{-13} seconds is taken as the periodic time of a molecular vibration and q is expressed in eV. Table 2.3 shows the value of τ at 295 K for a number of energy values q.

Table 2.3 Values of τ for various q at 295 K.

q (eV)	τ
0.2	3×10^{-10} s
0.4	1 μs
0.6	20 ms
0.9	400 s
1.1	1.2×10^6 s (=2 weeks)

Note that for values of q associated with weak binding (physical adsorption) the molecule's stay on the surface is fleeting, whereas for q greater than 1 eV the times are such that the molecule is very effectively trapped on the surface.

We have been discussing the state of an individual adsorbed molecule. The totality of such molecules forms an *adsorbed phase* on the solid surface but it will be clear that the population of adsorbed molecules is a changing one. Molecules arrive at the surface from the gas and are adsorbed; others desorb after a stay time on the surface whose average value is τ. Depending on the rate of arrival of molecules J, their probability of becoming adsorbed and the values of q and T which control the desorption process, the surface may be partially or completely covered with an adsorbed layer. Conditions may be such that there is multi-layer adsorption with subsequent layers being progressively more weakly bound. In the case of water vapour adsorption the state of binding in outer layers will approach that corresponding to condensation.

The probability of a molecule becoming adsorbed is expressed by coefficients appropriate to capture in the physical or chemisorbed states, and they depend on coverage. Initial coefficients on clean surfaces are generally high, in the range 0.1 - 0.5, but fall sharply as coverage increases. Even if a molecule does not become adsorbed the interaction with the surface is usually sufficiently prolonged that the molecule returns to the gas phase in a direction unrelated to that of its arrival. The importance of this is further discussed in section 2.6. The physics of the molecule - surface interaction is discussed by Redhead et al (1968) and recent advances are described by Somorjai (1994). An excellent introduction to the subject will be found in the text of Hudson (1992).

2.4.2 Outgassing

The broad relevance of the above to vacuum practice and outgassing is as follows. It should be stressed that in typical systems the condition of interior surfaces is not well known because of the variety and complexity of the processes which occur in the course of operating the system. In addition typical technical surfaces will be microscopically rough and oxidised, with many surface flaws.

When vacuum systems are vented back up to atmospheric pressure, direct adsorption from the gas phase leads to complete coverage of up to several layers of adsorbed gas. Large amounts of water vapour are adsorbed, as well as oxygen, nitrogen and other atmospheric gases. Some of this adsorbed gas will migrate into the interior near-surface region to become absorbed. It is prudent to try to minimise water vapour adsorption by venting to dry nitrogen, for example, because nitrogen is less strongly bound to structural materials such as stainless steel than is water vapour, and desorbs more readily in subsequent pumping down.

In pumping down from atmospheric pressure, most of the gas in the volume is soon removed, and, in typical systems, sub-millibar pressures of 10^{-2} mbar or better are achieved in times of the order of minutes. Desorbing gas starts to contribute to the gas load below about 10^{-1} mbar, and as the pressure continues to fall into the region below 10^{-4} mbar the gas load (assuming no leaks) becomes increasingly due to outgassing. Of the molecules which desorb, a number will find the entrance to the pump and be removed immediately. But others, a majority in vacuum chambers of typical proportions, will traverse the chamber to another part of the vacuum wall, and further interactions with the surface, before being pumped away. It is this traffic of molecules in transit to and fro across the chamber, fed by desorption and diminished by pumping, which constitutes the number density n of molecules in the vacuum and determines its quality.

Gas that is loosely bound on internal surfaces is pumped away quickly. Gas that is tightly bound desorbs at a very slow rate and does not contribute a significant load. But water vapour, which has an appreciable probability of desorption, has been stored in large amounts at the surface by adsorption during exposure to the atmosphere. The result is that there is protracted gassing of water vapour from structural materials such as stainless steel and glass. The outgassing rate does decrease, but only slowly and desorbing water vapour accounts for the dominant gas load for times of the order of tens of hours, unless bakeout procedures, to be discussed shortly, are used. In circumstances in which pump performance did not limit the vacuum achieved outgassing would be observed to diminish as time progressed, with the contribution to it of the small amount of diffusion of gas from the interior, particularly hydrogen, becoming more important and eventually, after very long times, becoming the dominant contribution.

In practicable experimental times therefore, which are of the order of hours, the main source of gas from unbaked surfaces is surface desorbed gas, principally water vapour. The outgassing rate may be quantified by a specific gassing rate q_G per cm^2 of surface. Typical values for well pre-cleaned surfaces of stainless steel, glass and ceramic, after a few hours pumping, are in the range 10^{-8} to10^{-9} mbar l s^{-1} per cm^2. This unit is introduced and explained in section 2.5. Tabulated outgassing rates as a function of time for a variety of representative materials are given, for example, by O'Hanlon (1989).

In high vacuum (HV) practice permeation of atmospheric gases through elastomer gaskets may, if there are many of them, contribute significantly to the ultimate gas load. Permeation through vessel walls is not significant in HV practice in comparison with other sources.

If UHV conditions, $p \sim 10^{-10}$ mbar and less are to be achieved outgassing has to be reduced by factors of 10^3 or more below the values which are achieved after a few hours pumping, and which are characteristic and limiting in the HV regime. This is done, once the pressure has reached the HV regime, by baking the whole system, vessel and contents, at temperatures of typically 200 °C or more for times of order 12 hours or more with continual pumping. All thermally activated processes are thereby greatly accelerated. Surface adsorbed gas is removed at a much greater rate than at room temperature, and gas absorbed in the bulk diffuses at a much greater rate to the interior surface where it desorbs and is pumped away. Both the surface and its hinterland in the vacuum wall are therefore depleted of gas molecules and, on cooling back to room temperature, the outgassing rate is dramatically reduced. Values of q_G are now typically in the range 10^{-12} to 10^{-13} mbar l s^{-1} per cm^2 and are consistent with the attainment of UHV pressures. Water vapour has been almost completely removed and the composition of the outgassing from stainless steel is dominantly hydrogen and carbon monoxide, which diffuse slowly from the interior. Hobson (1961) discusses outgassing, desorption energies and the role of bakeout in terms of a simple model which is very informative.

In UHV systems where pressures of 10^{-11} mbar or less are achieved, permeation of atmospheric hydrogen and helium through the vacuum wall at room temperature becomes significant, even though the concentration of these gases in the atmosphere is minute, of the order of a few ppm (parts per million). Hydrogen permeates stainless steel and helium permeates glass. The permeation of other gases is insignificant.

The improvement in the outgassing characteristics of materials continues to be a matter of concern and importance, as the proceedings of the 16th IUVSTA Workshop on the subject, held at Gräftåvallen in Sweden in 1996, will testify. Lower outgassing implies that better vacua may be achieved more quickly, a matter of concern, especially in an industrial context.

There is increasing understanding of outgassing processes as the analytical methods of surface science, originally developed to study semiconductor and other highly specialised surfaces, are brought to bear on the sorts of surface encountered in vacuum applications. This is exemplified by the study of Shin *et al* (1996) of the outgassing characteristics of stainless steel surfaces which have undergone various preparatory treatments.

Having described the various gas sources which present a load to the pump we next need to set up quantitative measures of gas flow.

2.5 GAS FLOW: FORMALITIES

2.5.1 Specifying quantity of gas in static circumstances

The quantity of a liquid may be most simply and directly specified by its volume, for example, one litre of whisky. This is a convenient measure of amount which, since liquids are virtually incompressible, is easily related via the density to the fundamental measure of quantity, namely the mass.   Gases however are compressible. The amount present in a given volume depends on the pressure. Think for example of a car tyre at various states of inflation in excess of what is normal, or the contents of a high pressure gas cylinder.

The fact that a given mass of gas may occupy a small volume at high pressure or, conversely, a large volume at low pressure presents a slight complication in specifying amount, but the problem is easily solved. Recall that $p = nkT$ with $n = N/V$. Then $pV = NkT$ where N is the number of molecules present in the volume V. Therefore at a given temperature, which may be taken as room temperature, 22 °C = 295 K, the product pV which is proportional to N gives a direct measure, apart from the constant factor kT, of the number of gas molecules in the volume. The product pV, expressed in millibar-litres (mbar l) can thus serve as a convenient measure of the amount of gas in static circumstances.

If it were necessary to specify the actual mass of gas this is easily done, for the mass of gas $= N \times m = (pV/kT)m$ (in consistent units).

Example 1. (a) Specify the amount of air in mbar l in an 'empty' one-litre bottle at room temperature and at a pressure of 1000 mbar. (b) How many such bottles would be needed to contain this gas at a pressure of 10^{-2} mbar?

Answers. (a) 1000 mbar × 1 litre = 10^3 mbar l.   (b) Let B be the number of bottles of volume 1 litre. By Boyle's law, $pV =$ constant. Therefore 10^3 mbar l $= 10^{-2} \times B$ mbar l. Whence $B = 10^5 = 100\ 000$, a lot of bottles.

Example 2. How many molecules are there in 1 mbar l of ideal gas at 22 °C (295 K)?

Answer. By equation (1.9), $N = pV/kT$ so that with $p = 1$ mbar $= 100$ Pa and $V = 1$ litre $= 10^{-3}$ m^3 and $kT = 4.07 \times 10^{-21}$ Joule, $N = 10^2 \times 10^{-3}/4.07 \times 10^{-21} = 2.46 \times 10^{19}$.

Example 3. To how many mbar l does 1 cm^2 of a monolayer of adsorbed gas (10^{15} molecules per cm^2) correspond?

Answer. $10^{15}/2.46 \times 10^{19} = 4 \times 10^{-5}$ mbar l, approximately.

This (pV) specification may be conveniently carried over to describe gas flow.

2.5.2 Gas flow and throughput, Q

Gases flow in response to pressure differences. Consider flow down the pipe in figure 2.4 when a pressure difference is maintained between its ends.

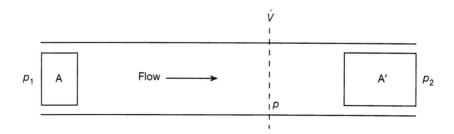

Figure 2.4 Gas flow down a pipe.

p_1 is greater than p_2 and so flow is in the direction shown. Because the volume occupied by a given mass of gas depends on its pressure a small volume labelled A near the inlet expands as it moves to occupy a volume A' near the outlet. The effect is exaggerated in the figure for emphasis, but must be accounted for in specifying flow. This is done as follows.

Pressure is constant across any particular cross-section and the flow may be specified as the product of the pressure p and the volume flow rate at this pressure through the cross-section. This is in accord with the pV description of static gas in the previous section. Denoting the volume flow rate at pressure p by $\dot{V}$ (the dot to emphasise that it is a volume per second) then we define throughput Q by

$$Q = p\dot{V} \quad \text{mbar l s}^{-1} \tag{2.4}$$

For steady flow Q is constant down the pipe (as much gas leaves it as enters) so that $p_1 V_1 = p_2 V_2$. Throughput Q is the basic quantity which specifies gas flow. It is easily related to the particle flow rate dN/dt since, using $pV = NkT$, with T fixed in this context and p constant at a particular cross-section

$$dN/dt = (d/dt)(pV/kT) = (p/kT)(dV/dt) = p\dot{V}/kT \qquad (2.5)$$

The particle flow rate is therefore

$$(dN/dt) = Q / kT \qquad (2.6)$$

and the mass flow rate $\dot{W}$, say, is

$$\dot{W} = m (dN/dt) = m Q/kT = MQ/RT \qquad (2.7)$$

where the last form results from multiplying by N_A in numerator and denominator.

Example. A fan moves atmospheric air through a room at a rate of 0.9 metres^3 per minute. What is the throughput in mbar l s^{-1}?
Answer. 1 m^3 = 1000 l so 0.9 m^3 = 900 l and $\dot{V}$ = 900/60 = 15 l s^{-1}. Therefore $Q = p\dot{V}$ = 15 × 1000 = 15 000 mbar l s^{-1}.

2.5.3 Speed, S

At certain places in vacuum systems, for example where gas enters a pipe from a vessel and also where it emerges from the pipe to enter the throat of a pump, it is usual to refer to the volumetric flow rates $\dot{V}$ as the speeds.
   As before, but with S replacing $\dot{V}$ in the formula $Q = p\dot{V}$

$$Q = Sp \qquad (2.8)$$

or

$$S = Q/p \ \text{l s}^{-1}$$

   Since it assumes a special role, we shall denote by S^* the speed at the inlet of a pump. In figure 2.5 a throughput Q is taken into a pump from a vessel at pressure p_1. At the pump $S^* = Q/p$ whilst at the vessel, where the pressure is higher, S has the smaller value $S = Q/p_1$. How S^* determines the upstream speed at the vessel is dealt with in section 2.5.5.

40   Basic Vacuum Technology

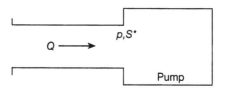

Figure 2.5 Gas flow into a pump.

2.5.4 Conductance, *C*

This measures ease of flow and is defined by

$$C = Q/(p_1 - p_2) \quad 1\,\text{s}^{-1} \tag{2.9}$$

where $(p_1 - p_2)$ is the pressure difference between two regions, say the inlet and outlet of a pipe, between which there is a throughput Q.

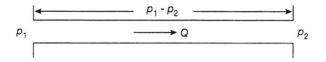

Figure 2.6 Defining conductance.

It is clearly a sensible definition - if the same pressure drop exists across a number of pipes of different sizes then that which allows the highest throughput has the greatest conductance.

Algebraic expressions and numerical examples for the conductance of various elements are given in sections 2.7 and 2.9.

Components may be connected in series or in parallel. For conductances connected in parallel the effective conductance of the combination is

$$C = C_1 + C_2 + C_3 \text{ etc} \quad \underline{\text{conductances in parallel}} \tag{2.10}$$

This may be deduced by reference to figure 2.7 in which just two conductances C_1 and C_2 connect regions of pressure p_1 and p_2 with $p_1 > p_2$.

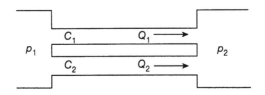

Figure 2.7 Conductances in parallel.

The separate throughputs are $Q_1 = C_1(p_1 - p_2)$ and $Q_2 = C_2(p_1 - p_2)$ so that the total throughput Q is

$$Q = Q_1 + Q_2 = (C_1 + C_2)(p_1 - p_2)$$

The effective combined conductance is therefore $C = C_1 + C_2$, and clearly the argument may be extended to any number of conductances in parallel.

For conductances connected in series the effective combined conductance C is given by

$$\frac{1}{C} = \frac{1}{C_1} + \frac{1}{C_2} + \frac{1}{C_3} + \text{etc} \quad \underline{\text{conductances in series}} \quad (2.11)$$

This may be proved by reference to figure 2.8 in which elements with conductances C_1, C_2 and C_3 are separated by volumes which enable intermediate equilibrium pressures p_2 and p_3 to be established.

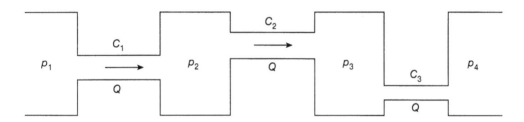

Figure 2.8 Conductances connected in series.

The steady state flow through the elements is

$$Q = C_1(p_1 - p_2) = C_2(p_2 - p_3) = C_3(p_3 - p_4) \text{ etc.}$$

Therefore

$$p_1 - p_2 = Q/C_1, \quad p_2 - p_3 = Q/C_2, \quad p_3 - p_4 = Q/C_3 \quad \text{etc.}$$

Adding these equations gives

$$p_1 - p_4 = Q(1/C_1 + 1/C_2 + 1/C_3)$$

The effective combined conductance C between the regions at p_1 and p_4 will be such that $Q = C(p_1 - p_4)$ so that by comparing these two expressions C is given by equation (2.11) above.

For just two conductances in series (2.11) becomes

$$C = \frac{C_1 \times C_2}{C_1 + C_2} \tag{2.12}$$

2.5.5 Pumping speed at the vessel

Consider a pump of speed $S*$ connected via a pipe of conductance C to a vessel where the pressure is p. Let S be the speed at the vessel and Q the throughput, as shown in figure 2.9.

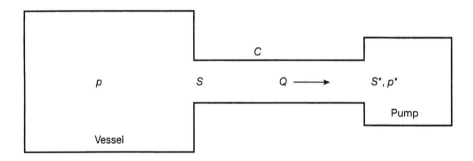

Figure 2.9 Effect of conductance on pumping speed.

The throughput Q is

$$Q = C(p - p*) = S \times p = S* \times p*$$

which gives after a little algebra

$$S = \frac{S* \times C}{S* + C} \tag{2.13}$$

This equation shows that the effect of the conductance is always to reduce the speed at the vessel. If the conductance C in $1\,s^{-1}$ is equal to the speed of the pump, the speed at the vessel is halved. Only with $C > 5S*$ is an effective speed approaching $S*$ realised.

Example.   Suppose a pump of speed $S* = 100\ 1\,s^{-1}$ is connected to a vessel by a pipe of conductance $400\ 1\,s^{-1}$. Find the speed (volumetric flow rate) at the vessel. If the pressure at the pump is 1.0×10^{-5} mbar what is the pressure in the vessel ?
Answer. From equation (2.13) the pumping speed at the vessel is $100 \times (400/500)$ $= 80\ 1\,s^{-1}$. The throughput into the pump is $Q = S* \times p* = 100 \times 10^{-5} = 10^{-3}$ mbar $1\,s^{-1}$.

As depicted in figure 2.9 the throughput entering the pipe from the vessel at pressure p is the same as that which goes into the pump. Therefore $80 \times p = 10^{-3}$ mbar l s^{-1} whence $p = Q/S = 10^{-3}/80 = 1.25 \times 10^{-5}$ mbar, compared with 1.0×10^{-5} mbar in the pump.

2.3.5 Summary

The two prime results for specifying gas flow are first

$$Q = S \times p$$

which relates throughput Q to volumetric flow rate S and the pressure at a particular place and secondly

$$Q = C(p_1 - p_2)$$

which determines flow between two regions at different pressures.

2.6 GAS FLOW: MECHANISMS

How gas flows depends on its state, as determined by the Knudsen number Kn introduced in section 1.12. For small Knudsen numbers, Kn < 0.01, gas is in its familiar fluidic (continuum) state with inter-molecular collisions dominating its behaviour. For Kn > 1, gas is in a molecular state in which inter-molecular collisions are negligible and molecules collide only with the containing surface, of a pipe for example. For the intermediate range of Kn between 0.01 and 1 gas is described as being in a transitional state, neither continuum nor molecular but with some of the attributes of both. With each of these gas conditions is associated a different character of flow and Kn therefore determines the *flow regime* as being *continuum, transitional* or *molecular*.

2.6.1 Continuum flow

To fix our ideas let us consider a pipe of diameter $D = 5$ cm (2 inch). As table 2.2 shows at a pressure 10^{-3} mbar the mean free path is 6.6 cm which is about the same as the pipe diameter. The mean free path will therefore be 100 times smaller than this at a pressure 10^{-1} mbar. Therefore at 10^{-1} mbar and all higher pressures gas behaviour will be in the continuum regime and fluidlike. Gas can be pushed along the pipe and sucked out of it. Movement of the gas, once induced, is communicated by inter-molecular collisions.

Depending on how large is the pressure difference driving the flow in this continuum state, it is either *turbulent* or *viscous*. In turbulent flow, which is associated with relatively large pressure differences, the motion of the gas in the general direction of the pressure difference is accompanied by irregular and unpredictable local fluid movements in the form of swirls and eddies. In viscous flow, the motion of the gas is more orderly, with stable, well defined flow patterns which are controlled by its viscosity. Whether flow is turbulent or viscous is determined by the value of the dimensionless Reynolds' number Re defined as

$$Re = \frac{\rho u D}{\eta}$$

where ρ is the gas density, u the velocity of the flow, D the pipe diameter and η the viscosity coefficient. For Re > 2000, flow is turbulent; for Re < 2000 it is viscous.

As is shown in appendix C this criterion can be expressed in more useful vacuum-related quantities involving the throughput Q and pipe diameter D. For air at room temperature the condition for turbulence then becomes

$$Q/D > 244, \quad Q \text{ in mbar l s}^{-1}, \ D \text{ in cm, \quad for turbulence}$$

Thus for $D = 5$ cm, throughputs greater than 1220 mbar l s^{-1} are turbulent. This is a relatively high throughput. A flow Q of 1500 mbar l s^{-1} in this pipe at a pressure 50 mbar implies a volumetric flow rate of 30 l s^{-1} and a high associated gas velocity u ~ 15 m s^{-1}. Such conditions are encountered in some industrial processes, but relatively rarely, and in most applications of vacuum technology it is viscous flow which is encountered. Nevertheless abrupt changes in flow conditions caused by a restriction or sharp bend in a pipe, or the sudden opening of a valve for example, may cause local or transient turbulence respectively, after which viscous conditions are restored.

The analysis of viscous flow conditions in this continuum flow regime - that is, relating a throughput to a pressure difference for a given component, typically a pipe - is in general not straightforward. In most applications of vacuum technology the concern is to achieve the best performance from pumps by choosing connecting pipes which are as short and wide-bored as is practicable. The formula of Poiseuille for gas flow down a pipe is only accurate for very long pipes with L/D of order 100 and is therefore restricted in its application. It gives the conductance C for air at room temperature of a pipe of length L and diameter D as

$$C = \frac{136D^4}{L}\left(\frac{p_1 + p_2}{2}\right) \quad 1\,\text{s}^{-1} \quad D, L \text{ in cm} \qquad (2.14)$$

where the bracketed term is the average of the inlet and outlet pressures. If used in circumstances where $L/D \sim 10$ this formula is likely to be in error by several tens of percent. The problems associated with accurate flow prediction and analysis in this flow regime, and their solution, are described in a definitive article by Livesey (1998). As is there emphasised, the prime purpose of analysis is to relate throughput to the pressure conditions. Whilst conductance C formally does that, there is sometimes other information available which enables the flow problem to be solved and the quantities required to be determined. In other words, determining the conductance may not be of prime importance. The reader who needs to take these matters further should consult this source.

For applications in which this continuum pressure regime and the transitional regime are 'passed through' on the way down to high vacuum and molecular conditions precise knowledge of flow is fortunately not likely to be important.

2.6.2 Molecular flow
Consider again the pipe of 5 cm diameter referred to above. For pressures of 10^{-3} mbar and less, the mean free path is <u>greater</u> than the pipe diameter and gas conditions are molecular. This has profound significance for the way in which gas flows, and it is quite different in character from continuum flow. Molecules travel from wall to wall across the vacuum space without colliding with each other, as depicted in figure 2.10.

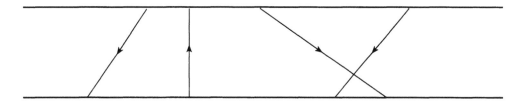

Figure 2.10 Molecular trajectories in molecular flow conditions.

There is no fluid behaviour in the conventional sense - the gas cannot be pushed along the pipe as in viscous flow because there is no means of communicating motion between the molecules of the gas. The gas exists as a very large number of completely independent, non-interacting molecules. Let us examine the motion of molecules in more detail.

As discussed under 'Surface processes' when molecules reach a wall they may be temporarily adsorbed there before being released by desorption. The direction of departure of the desorbing molecule is random - it retains no 'memory' of its original direction of arrival at the surface. Molecules which spend some time on the surface without becoming adsorbed return to the gas in a random direction as previously noted. Even if, with very small probability, a molecule does rebound directly back to the vacuum it does so from a real surface which is rough on a microscopic scale, so that there is no simple 'law of reflection' against a smooth surface such as operates when a snooker ball rebounds from the edge of the table. The result is that, whatever the direction of arrival of a molecule at the vacuum wall, it returns to the vacuum in a random direction unrelated to its direction of arrival. Essentially, the molecule's motion is randomised by its interaction with the surface.

This is the state of affairs described by Knudsen's cosine law, according to which the probability of a molecule being scattered from the surface in a particular direction is proportional to cosine θ where θ is the angle which the direction of departure makes with the normal to the surface. See figure 2.11.

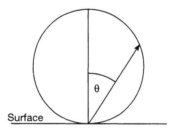

Surface

Figure 2.11 Illustrating the cosine law.

Walton (1983) may be consulted for a brief general discussion and Steckelmacher (1986) for a discussion of the law's particular relevance to vacuum practice. See appendix D for more discussion.

Returning to the matter of flow, we imagine as in figure 2.12 a particular molecule which is 'released' as shown at the centre of a long pipe. Its subsequent trajectory will be such that, on average, it tends to return to the locality where it started. There is no preference for motion to the left or right, and no way in which the independent molecule's motion can be influenced.

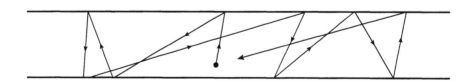

Figure 2.12 Random to and fro motion of a molecule in a long pipe.

Flow is possible in these circumstances only because, as depicted in figure 2.13, a suitable pump connected to one end of a pipe will capture a fraction of the molecules which enter it and prevent their return down the pipe.

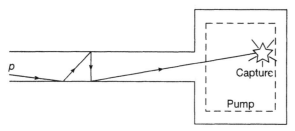

Figure 2.13 Capture of a wandering molecule by a pump.

By capturing molecules and removing them from the gas phase the pump will maintain a lower number density of molecules at its inlet than exists at the other end of the pipe where there is a higher number density n corresponding to the pressure p in the vessel. Overall therefore there is a concentration gradient of molecules along the pipe and a steady net diffusive motion from a region of higher density at the vessel to the pump where a lower density is maintained by continuous capture due to the pumping action. If pumping ceases the density variation disappears as molecules fill up the space to a uniform density.

An extremely important aspect of this molecular flow process is that the pump cannot 'suck' molecules along the pipe. All it can do is attempt to capture them when they arrive. The arrival rate of molecules at the pump inlet is determined by the gas alone, and this, as we will show, places an upper limit on the speed of any pump which operates in the molecular flow region.

2.6.3 The transitional regime

Figure 2.14 shows how the conductance of a pipe changes from viscous to molecular in character through the transition regime as pressure is reduced. As previously noted, this regime is characterised by values of the Knudsen number Kn between 10^{-2} and 1. Flow in this regime is difficult to analyse and is described most simply by a combination of the two flows with an appropriate combination factor. See Lewin (1965) and Livesey (1998).

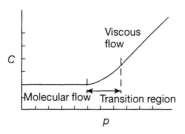

Figure 2.14 The various flow regimes.

The main feature to note from the figure is that conductance is independent of pressure in the molecular flow range, whereas for viscous flow it increases with pressure. Fundamentally this is because in the viscous flow range, as noted, force may be communicated to a gas to make it move. In molecular flow by contrast the molecules are independent. There is no way of communicating with them to make them go in a particular direction.

2.7 MOLECULAR FLOW CONDUCTANCE OF AN APERTURE

Consider an aperture of area A separating two regions maintained at different pressures p_1 and p_2, with $p_1 > p_2$ as shown in figure 2.15.

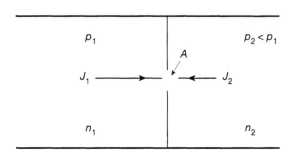

Figure 2.15 Molecular flow at an aperture.

Corresponding to p_1 and p_2 will be fluxes J_1 and J_2 with $J_1 > J_2$ as indicated, so that there will be a net flow of molecules from left to right given by $dN/dt = (J_1 - J_2) A$. Substituting for the J from equation (1.7) and using equation 2.6 ($Q = kT(dN/dt)$) to convert from a number rate to a throughput specification gives a net left-to-right throughput

$$Q = \sqrt{\frac{kT}{2\pi m}} A(p_1 - p_2) = \sqrt{\frac{RT}{2\pi M}} A(p_1 - p_2)$$

so that the conductance of an aperture, which we shall denote as C_0, is

$$C_0 = \sqrt{\frac{RT}{2\pi M}} A \qquad (2.15)$$

This is an important result. Note the presence of the facto $\sqrt{T/M}$. For nitrogen at 295 K it gives

$$C_0 = 11.8\,A\ \mathrm{l\,s^{-1}}, \ A \text{ in cm}^2 \qquad (2.16)$$

For a circular aperture of diameter D cm it takes the useful form

$$C = 9.3D^2 \quad \text{l s}^{-1}, \ D \text{ in cm} \tag{2.17}$$

Since this is molecular flow, with no molecule - molecule collisions, the flows in each direction are quite independent and do not interfere. They are determined solely by the conditions on the side from which they originate. On each side therefore, it is as though there are no molecules in the region beyond and we can express the left-to-right and right-to-left throughputs at the plane of the aperture as $Q_1 = C(p_1\text{-}0) = Cp_1$ and $Q_2 = C(p_2\text{-}0) = Cp_2$ respectively, thus giving $Q = Q_1 - Q_2 = C(p_1 - p_2)$ correctly.

2.8   MAXIMUM SPEED OF A PUMP IN THE MOLECULAR FLOW REGION

Consider, as in figure 2.16, a pump of inlet diameter D which is perfect in its action, so that its mechanism captures and suitably disposes of all molecules which enter it, and there is no return flux (corresponding to $J_2 = 0$ in figure 2.15).

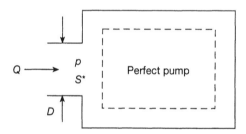

Figure 2.16 Action of an ideal pump.

At the inlet of the pump the throughput is $Q = S^*p$ which can also be expressed, in the light of the comments at the end of the last section, as $Q = Cp$ where C is the conductance of the aperture which forms the inlet to the pump, sometimes referred to as its throat. Hence in the case of an ideal pump, its speed is equal to the conductance of its inlet aperture. That is

$$S^* = C_{inlet} \ \text{l s}^{-1} = 9.3 \, D^2 \ \text{l s}^{-1} \quad D \text{ in cm}$$

A pump with a 10 cm (4 inch) inlet diameter would therefore have a speed of 930 l s^{-1} if it worked ideally. Pumps do not work ideally and do not capture all the arriving molecules. Therefore $S^* < C_{inlet}$ and measured speeds S^* reflect this, S^* being the net captured flow into the pump. The ratio (S^*/C_0) is therefore a measure of pump efficiency. Pumps will be discussed in chapter 3.

2.9 MOLECULAR FLOW THROUGH PIPES; TRANSMISSION PROBABILITY AND CONDUCTANCE

2.9.1 General considerations

Consider as in figure 2.17 a pipe of length L, diameter D and cross-sectional area A connecting two regions of low pressure p_1 and p_2 such that $\lambda \gg L$, D and conditions are those of molecular flow, as previously discussed in section 2.6.2.

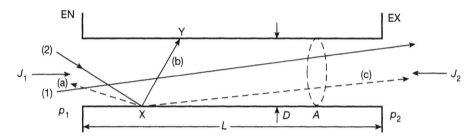

Figure 2.17   Molecular flow through a pipe.

The total number of molecules per second crossing the plane EN to enter the pipe is J_1A. They come from all directions within the left-hand volume. Relatively few, like molecule (1), will be travelling in such a direction as to pass right through the pipe without touching the sides, but most will not. The majority, like molecule (2), will collide with the wall at a place such as X and return to the vacuum in a random direction as discussed in section 2.6.2. There are now three possible outcomes (a), (b) or (c) as shown by the dashed trajectories. The molecule may (a) return to the left-hand volume, (b) go across the pipe to Y - and then another 'three-outcome' event - or (c) leave the pipe through the exit plane EX into the right-hand volume. These three outcomes occur with different probabilities, each of which has two aspects - the probability of take-off in a particular direction as determined by the cosine law (section 2.6.2) and whether that direction is within the solid angle subtended at X by the entrance area, the exit area, or the rest of the pipe. Furthermore for a molecule which goes to Y at a different distance along the pipe the balance of probabilities for where it next goes will have changed accordingly. A little reflection shows that this is quite a complex problem to analyse, not least because of its three-dimensional character which is disguised in a two-dimensional diagram.

Considering the total flux of molecules J_1A which enter the pipe at EN, and the diverse possibilities for their subsequent trajectories, it is clear that some molecules will eventually be transmitted through the exit plane EX. The remainder will return back through the plane EN. The fraction which do pass through EX into the right-hand region may be defined as the transmission probability α of the pipe so that the transmitted flux is $\alpha(J_1A)$. It is clear that α will be large for short pipes and for $L \ll D$ will approach unity, corresponding to the flow through an aperture.

Increase of L, because of the increased number of wall interactions, must cause α to decrease.

For flow in the right-to-left direction similar considerations must apply. The transmission probability of the pipe must be the same in both directions, but the flux J_2 corresponds to the lower pressure p_2. The right-to-left flow is therefore $\alpha(J_2 A)$ and the net observable flow is the difference of the flows in each direction. Thus, multiplying this net flow rate $\alpha(J_1-J_2)A$ by kT to get a throughput specification gives

$$Q = kT(J_1 - J_2)A\alpha$$

Substituting for J from equation (1.7) leads to

$$Q = \sqrt{\frac{kT}{2\pi m}}\, Aa(p_1 - p_2)$$

$$= \sqrt{\frac{RT}{2\pi M}}\, Aa(p_1 - p_2) \tag{2.18}$$

in which we recognise from equation (2.15) the aperture conductance C_0 so that

$$Q = \alpha\, C_0\, (p_1 - p_2)$$

The conductance of the pipe may therefore be expressed in terms of its entrance conductance C_0 and transmission probability α as

$$C_{\text{pipe}} = \alpha C_0 \tag{2.19}$$

The concept of transmission probability can be widely and usefully applied, as will become evident in subsequent sections.

2.9.2 Analysis

An approximate but useful working expression for the conductance of a pipe of arbitrary length is given on page 52 in equation (2.24). It is however necessary to comment on the status of this formula, because, as may be gathered from the discussion above, the problem is not straightforward.

There have been several different approaches made to the calculation of the molecular flow conductance of a pipe. The pioneering work of Knudsen in 1909, based on an assumption about molecular drift velocities and the application of the principles of mechanics, was a strong stimulus to other workers in the following three decades, notably Smoluchowski, Dushman and Clausing, who developed the transmission probability method. More recently Steckelmacher made important contributions to the subject. His review of 1986 on the current state of the art should be consulted, together with an earlier one of 1966.

For very long pipes such that $L \gg D$ the conductance C_L calculated by the Knudsen method is correctly given to a 1% order of accuracy by

$$C_L = \frac{D^3}{6L} \sqrt{\frac{2\pi RT}{M}} \qquad (2.20)$$

A derivation of this result is given in appendix E. Again, as with the equation for C_0, the presence of the factor $\sqrt{T/M}$ should be noted. Once numerical expressions are established for nitrogen, the conductance for other gases may easily be obtained. For nitrogen at 295 K, expressed in litres per second, it becomes

$$C_L = 12.4 D^3 / L \ \ 1 \, \text{s}^{-1}, \quad D, L \text{ in cm} \qquad (2.21)$$

For the pipes usually encountered in vacuum practice, which are not long compared with their diameter, Dushman suggested that their conductance be calculated as that of an aperture conductance C_0 appropriate to the entrance area in series with a pipe conductance given by the long pipe expression above. Thus for a pipe of any length, using equation (2.12)

$$C_{\text{pipe}} = \frac{C_0 \times C_L}{C_0 + C_L} \qquad (2.22)$$

From equations (2.17) and (2.21) above

$$C_L / C_0 = 4D/3L$$

Rearranging equation (2.22) and substituting for C_L / C_0 gives

$$C_{\text{pipe}} = \frac{C_L}{1 + C_L / C_0} = \frac{C_L}{1 + 4D/3L} \qquad (2.23)$$

Thus for nitrogen gas at 295 K, using equation (2.21)

$$C_{\text{pipe}} = \frac{12.4 D^3 / L}{1 + 4D/3L} \ \ 1 \, \text{s}^{-1}, \quad D, L \text{ in cm.} \qquad (2.24)$$

This is the useful working formula referred to above and it is sufficiently accurate for most purposes. It may be in error, due to the method of its derivation by amounts up to 10%, depending on the particular L/D ratio in question. If more accurate values are required, Clausing's data quoted by O'Hanlon (1989) should be used in conjunction with equation (2.19).

In equation (2.24) above the dependence of conductance on D^3 should be noted. For the highest conductance connecting pipes should be as short and fat as design constraints allow.

Finally, it is worthwhile to recast equation (2.22) into a different form to illustrate pipe conductance from the point of view of transmission probability. It may be rearranged to give

$$C_{\text{pipe}} = \frac{C_0}{1 + C_0/C_L} = \frac{C_0}{1 + 3L/4D}$$

so that, since from equation (2.19) namely $C_{\text{pipe}} = \alpha C_0$

$$\alpha = \frac{1}{1 + 3L/4D} \tag{2.25}$$

This shows that as L approaches zero, α correctly approaches unity, and diminishes with increasing L. Of course, the same remarks made above concerning accuracy apply. For nitrogen gas at 295 K

$$C_{\text{pipe}} = \frac{9.3D^2}{1 + 3L/4D} \quad 1\,\text{s}^{-1}, \quad D, L \text{ in cm} \tag{2.26}$$

which is equivalent to equation (2.24).

Example 1. Calculate the molecular flow conductances for nitrogen at 295 K of:
(a) a circular aperture of diameter 5 cm
(b) a pipe of diameter 5 cm and length 20 cm
(c) a pipe of the same diameter as in (b) but 1 m long
(d) a pipe of the same length as (c) but 2.5 cm diameter.

Answers.   Using equations (2.17) or (2.24) as appropriate
(a) $C_0 = 232\,1\,\text{s}^{-1}$
(b) $C = 58\,1\,\text{s}^{-1}$
(c) $C = 14.5\,1\,\text{s}^{-1}$
(d) $C = 1.9\,1\,\text{s}^{-1}$.
Notice the marked reduction in conductance comparing (a), (b) and (c), and the effect of halving the diameter between examples (c) and (d).

Example 2.   For (b) and (c) in example 1 use the aperture result (a) and the Clausing factors from O'Hanlon (1989) to obtain more accurate estimates.

Answer.   The aperture conductance is 232 l s^{-1} and the L/D values are 4 and 20 respectively. From O'Hanlon p 36, $\alpha = 0.2316$ and 0.0613 respectively. Therefore for the 20 cm pipe, $C = 232 \times 0.2316 = 54$ l s^{-1} and for the 1 m pipe $C = 232 \times 0.0613 = 14.2$ l s^{-1}. Comparison with the values obtained above suggests that, for many purposes, the values obtained from formula (2.24) will be sufficiently accurate.

Example 3.   For the aperture and 20 cm long pipe in example 1 calculate the conductances for hydrogen at 295 K.

Answer.   As noted the molecular flow conductances involve the factor $\sqrt{T/M}$. Therefore $C_{hydrogen}/C_{nitrogen} = \sqrt{28/2} = 3.74$, whence the aperture and pipe conductances for hydrogen are $232 \times 3.74 = 868$ l s^{-1} and $58 \times 3.74 = 217$ l s^{-1} respectively.

Example 4.   What is the length-to-diameter ratio for a pipe whose transmission probability is 0.5?

Answer.   From equation (2.25), $\alpha = 0.5$ for $3L/4D = 1$, i.e. for $L = 1.3D$. Using Levenson's data quoted by O'Hanlon the ratio is more accurately deduced as $L = 1.1D$. However the point to be made   is that even for a very short pipe whose length is equal to its diameter, the transmission probability and therefore the conductance is approximately halved compared with that of an aperture of the same area.

2.9.3 Conductance of complex structures

The calculation of the conductance of complex shapes such as right-angled elbows and traps has followed the transmission probability approach, in which statistical Monte Carlo methods have proved very powerful. The important case of the elbow was worked out by Davis (1960). It emerges, perhaps surprisingly at first sight, but not when one reflects on the processes involved, that the conductance in molecular flow of an elbow is little smaller, typically by only a few per cent, than that of a straight tube of the same diameter and centre length. A useful compilation of the results of various workers, which includes data for rectangular and annular ducts, elbows and chevron baffles, is given by O'Hanlon (1989). The results of these analyses are presented in a useful graphical and parametrised form. Thus, for example, for a rectangular duct of section dimensions a and b with length L, data are given in terms of the ratios a/b and L/b. The appropriate value of α is determined from the graph   and then the conductance of the duct is found by multiplying this by the entrance conductance, in this case $11.8(a \times b)$ l s^{-1} from equation (2.16).

If a compound conductance problem seems to be susceptible to analysis by the series method using equation (2.11) caution should be exercised, as this strictly applies only for random entry conditions which are assured in the associated figure 2.8 by the presence of large intervening volumes between the elements. Elements which are directly connected in series are more properly dealt with by probabilistic methods developed by Oatley, which are concisely described and illustrated in the articles by Carlson (1979) and Livesey (1998). The inappropriate use of equation (2.11) is likely to give large (30% or more) errors. Accurate values of transmission probability obtained by Monte Carlo methods for pipes of different diameter connected in series are given by Santeler and Boeckmann (1987).

2.10 QUANTITATIVE DESCRIPTION OF THE PUMPING PROCESS

We are now in a position to describe the overall performance of a vacuum system quantitatively. Figure 2.18 depicts a vessel of volume v connected by a pipe of conductance C to a pump of speed S^*. The speed at the vessel is $S = S^* \times C/(S^* + C)$ from equation (2.13).

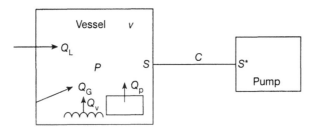

Figure 2.18 Quantitative description of the pumping process.

Gas is launched into the vessel from various sources. Denoting the total gas load by Q_T

$$Q_T = Q_G + Q_L + Q_V + Q_P \qquad (2.27)$$

In a particular application, one type of source is likely to be dominant. Good engineering and clean working practices will minimise or even eliminate Q_L and Q_V respectively. There may or may not be a process load Q_P depending on the application.

Consider the state when pressure is falling and before a steady state has been reached. Let p be the pressure in the volume v of the vessel at a given instant and suppose that it falls to a value $p - dp$ in a small time interval dt. The following word equation completely describes the system: the change in the amount of gas in the vessel in the time interval dt is equal to the amount removed by the pump minus the amount entering from all sources. In symbols

$$- v \, dp \; = \; Sp \, dt - Q_T \, dt$$

or
$$v \, (dp/dt) \; = \; - Sp \; + \; Q_T \qquad\qquad (2.28)$$

This is the basic equation of flow. It may be used to estimate the vacua achieved and pump-down times. When a system reaches steady state, the removal rate of gas from the volume is equal to the influx rate, $Sp_u = Q_T$. This is the dynamic balance referred to earlier and the pressure achieved is the ultimate pressure, p_u, under the prevailing conditions. Directly then, or putting $dp/dt = 0$ in equation (2.28)

$$p_u = Q_T \, / \, S \qquad\qquad (2.29)$$

This important equation, as well as determining the vacuum achieved tells us what is in one sense obvious - that the best vacua are achieved for small loads Q_T and large speeds S. However, as we have seen, S cannot be made arbitrarily large but is limited by a pump's throat conductance. The equation therefore emphasises the importance of reducing gas load as the only means of achieving better vacua once the optimum choice of pump has been made.

Recognising that the residual gas will probably consist of various gases A, B, C ... for which the pump may have different speeds S_A, S_B, S_C ... we may modify equation (2.29) to read

$$p_u = p_{u,A} + p_{u,B} + ... = Q_A/S_A + Q_B/S_B + ... \qquad\qquad (2.30)$$

where p_u is the sum of ultimate partial pressures.

Equation (2.28) can be used to tell us how pressure varies with time under various conditions.

Consider the early stages of pump-down from atmospheric pressure using a roughing pump. Assume no leaks, $Q_L = 0$, so that $Q_T = Q_G$ is negligible compared with the gas in the volume and can be put equal to zero. With $Q_T = 0$ equation (2.28) becomes, after rearrangement

$$dp/p = -(S/v) \, dt$$

This may be integrated to yield
$$p = p_0 \exp\{-t \, / \, (v \, /S)\} \qquad\qquad (2.31)$$

showing how pressure falls from an initial value p_0 in a vessel of volume v with a pump speed S, assumed constant. Note that the quantity (v/S) serves as a time constant T for the system so that $p = p_0 \exp(-t/T)$ where $T = (v/S)$. Since $e^3 = 20$, $e^{-3} = 1/20$ and equation (2.31) shows that the pressure will have fallen to 1/20 of its initial value after a time 3T.

Equation (2.31) may be re-expressed as

$$t = (v/S) \ln(p_0/p) \tag{2.32}$$

so that the time necessary for the pressure to fall from p_0 to p may be determined. For example if $v = 40\ l$ and $S = 0.5\ l\ s^{-1}$ then the time taken for the pressure to fall from $p_0 = 1000$ mbar to 1 mbar is

$$t = (40/0.5)\ln 10^3 = 80 \times 2.3\log_{10}10^3 = 80 \times 2.3 \times 3 = 552\ s \sim 9\ min$$

Equation (2.32) may also be used to determine the pumping speed necessary to rough down a volume to a given pressure in a specified time. By slight rearrangement

$$S = (v/t) \ln(p_0/p) \tag{2.33}$$

If a volume of $1\ m^3 = 10^3\ l$ has to be pumped down from 1000 mbar to 10 mbar in 5 min = 300 s then

$$S = (1000/300)\ln(10^2) = 3.3 \times 2.3 \times 2 = 15\ l\ s^{-1} = 900\ l\ min^{-1} = 54\ m^3 h^{-1}$$

Remember that these may be approximate estimates, erring on the optimistic side because in this pressure regime the viscous flow conductance of any connection to the pump is a function of pressure, so that the speed S at the vessel used above is not strictly constant. Figure 2.19 illustrates p as a function of time.

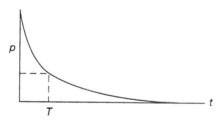

Figure 2.19 Illustrating equation (2.31).

Consider now a situation where Q_T can not be ignored. Suppose there is a relatively large leak such that Q_L dominates other sources and we may take $Q_T = Q_L$. The system will by equation (2.29) come to an ultimate pressure $p_{u,L}$ given by

$$p_{u,L} = Q_L/S \tag{2.34}$$

Consider finally the important situation of a leak free system where $Q_T = Q_G$. This is the state of affairs in HV and UHV systems which are working properly. The ultimate pressure achieved is p_u and by equation (2.29)

$$p_u = Q_G/S$$

As discussed earlier, the achievement of an ultimate pressure in the UHV range depends on reducing the gassing load substantially below HV values, by bakeout.

In either the HV or UHV regimes the Q_G value in the later stages of pumping decreases very very slowly, as noted in section 2.5, so that it may be taken as constant with an ultimate pressure value given by equation (2.34):.

If an influx of gas raises the system pressure from p_u to p_1 and the speed may be considered constant, then substituting $Q_G = Sp_u$ from equation (2.34) in equation (2.28) gives

$$v \, (dp/dt) = -S(p - p_u)$$

which integrates to give

$$p = p_u + (p_1 - p_u) \exp\{-t \, /(v/S)\}$$

so that the system pressure should recover on a time constant $T = v/S$ to its former value p_u.

A numerical example of the analysis of gas flow in a simple system is given in appendix F.

This chapter has dealt with vacuum systems in which the vacuum chamber is a large vessel within which the pressure may be regarded as uniform. Such systems comprise the vast majority encountered. The less frequently met, but nevertheless important systems which consist of a pump connected to a 'distributed' volume are discussed in appendix G.

Summary
We have now discussed the origins of the gas load in vacuum systems, how the gas flows and how the flow may be described quantitatively. The next matters to deal with are how vacua are created and measured in practice.

References
Berman A 1996 *Vacuum* **47** 327
Carlson R W 1979 *Methods of Experimental Physics* vol 14, ed G L Weissler
        and R W Carlson (New York: Academic)
Davis D H 1960 *J. Appl. Phys.* **31** 1169
De Boer J H 1953 *The Dynamical Character of Adsorption* (Oxford: Clarendon)

Elsey J H 1975 *Vacuum* **25** 299 - 306 and 347-61

Hobson J P 1961 *Trans. 8th Nat. Vac. Symp.* vol **1** (New York: Pergamon) 26

Hudson J B 1992 *An Introduction to Surface Science* (Boston: Butterworth - Heinemann)

Knudsen M 1909 *Ann. Phys. Lpz.* **28** 75 - 130

Lewin G 1965 *Fundamentals of Vacuum Science and Technology* (New York: McGraw-Hill)

Livesey R G 1998 *Foundations of Vacuum Science and Technology* ed J Lafferty (New York: Wiley)

O'Hanlon J F 1989 *A User's Guide to Vacuum Technology* 2nd edition (New York: Wiley)

Redhead P A, Hobson J P and Kornelsen E V 1968 *The Physical Basis of Ultra-high Vacuum* (London: Chapman and Hall)

Roth A 1990 *Vacuum Technology* 3rd edition (Amsterdam: Elsevier Science)

Santeler D J and Boeckmann M D 1987 *J. Vac. Sci. Technol.* A5 2493 - 2496

Shin Y H, Lee K J and Chung Jhung K H 1996 *Vacuum* **47** 679

Somorjai G A 1994 *Introduction to Surface Chemistry and Catalysis* (New York: Wiley)

Steckelmacher W 1966 *Vacuum* **16** 561 - 583

Steckelmacher W 1986 *Rep. Prog. Phys.* **49** 1083 - 1107

Walton A J 1983 *The Three Phases of Matter* 2nd edition (Oxford: Clarendon)

3

Pumps

B S Halliday

A vacuum pump is a device for creating, improving and/or maintaining a vacuum. Two basic categories exist, the gas transfer and the entrapment groups. The gas transfer group can be subdivided into positive displacement pumps in which repeated volumes of gas are transferred from the inlet to the outlet usually with some compression, and the kinetic pumps in which momentum is imparted to the gas molecules so that gas is continuously transferred from the inlet to the outlet. In contrast, entrapment pumps are those which retain molecules by sorption or condensation on internal surfaces. Table 3.1 shows the various pumps in their appropriate groups which are described later, gas transfer pumps in sections 3.1 and 3.2 and entrapment pumps in section 3.3.

Table 3.1 shows the categories of pumps, but it must be considered with the pressure ranges over which these pumps work. These are shown in table 3.6 at the end of this chapter.

Table 3.1 Classification of vacuum pumps.

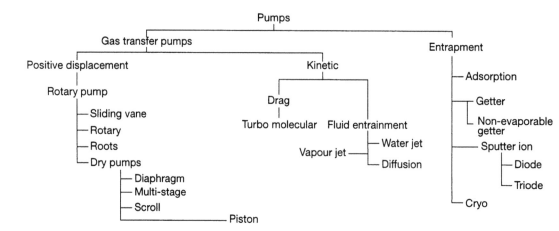

3.1 POSITIVE DISPLACEMENT PUMPS

These are pumps in which a volume filled with gas is cyclically isolated from the inlet, the gas being then transferred to the outlet, and usually compressed before discharge.

3.1.1 The rotary vane pump

This is probably the pump most widely used and is well established as a primary pump and also as a backing/roughing pump in many compound systems (refer to chapter 8).

An eccentrically placed slotted rotor turns in a cylindrical stator (figure 3.1) driven by a directly coupled electric motor. In the slots are two (or three) sliding vanes which are in continuous contact with the walls of the stator. Air (or gas) is drawn in, compressed and expelled through a spring loaded exhaust valve. The vanes and rotor are sealed by a fluid film, with the stator immersed in the fluid to provide heat transfer to the pump casing. Rotary pump fluids are described in section 5.5.

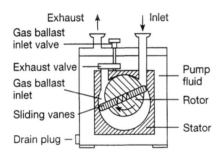

Figure 3.1 The sliding vane rotary pump.

Two-stage pumps are available where the exhaust from the first stage is internally connected to the inlet of the second stage (figure 3.2). This improves the ultimate pressure of the pump by reducing the back leakage where the rotor and stator are fluid sealed.

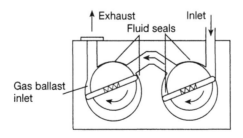

Figure 3.2 A cross-sectional diagram of a two-stage sliding vane pump.

Gas ballasting is used to reduce the extent of condensation of vapours during the compression cycle. Gas ballasting can be used where a controlled quantity of a suitable non-condensable gas (usually air) is admitted during compression (figure 3.3).

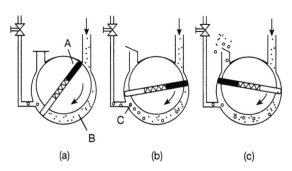

(a)   (b)   (c)

Figure 3.3 The gas ballast operating in a sliding vane pump.

In figure 3.3(a) vane A is about to close the crescent shaped chamber B containing the gases and condensable vapours being pumped. After this volume is sealed (b) from the inlet, a controlled volume of air at atmospheric pressure is admitted at C. This raises the pressure in B and prevents condensation by opening the exhaust valve at an earlier stage so that condensation conditions are not reached. Figure 3.4 shows the relative performance curves of single- and two-stage pumps operating with and without gas ballasting.

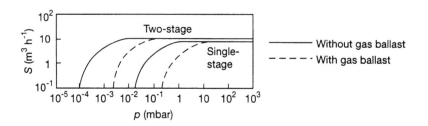

Figure 3.4 Typical pumping speeds for a sliding vane pump.

3.1.2 The rotary plunger pump

This pump is chiefly used for large volumetric displacement and has an eccentrically mounted rotor in a circular stator. The stator chamber is divided into two parts of varying volume by a vane rigidly fixed to the rotor. The vane slides in a plug oscillating in an appropriate housing in the stator. This pump can also be used with gas ballast. The plunger pump can be two-stage and can have speeds of up to 1500 $m^3 h^{-1}$, the larger pumps being water cooled. The ultimate pressure is about 10^{-2} mbar.

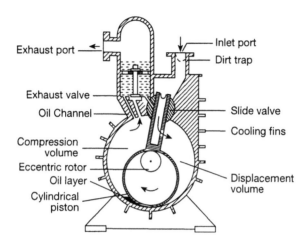

Figure 3.5 Section of a plunger pump.

3.1.3 Roots pump

The Roots pump shown schematically in figure 3.6(a) is used in the low and medium vacuum ranges with high volumetric flow. It cannot exhaust to atmospheric pressure as explained below and is operated in series with a backing pump, often of the rotary vane variety, or 'dry' pump, connected to the exhaust, as in figure 3.6(c). The Roots pump has two lobed rotors interlocked and synchronised by external gearing which rotate in opposite directions, moving past each other and the chamber with small clearance (0.3 mm). The gas is displaced from the inlet to the outlet. A typical performance curve is shown in figure 3.6(b).

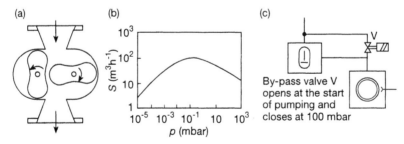

Figure 3.6 (a) Roots pump. (b) Typical performance curve. (c) Diagram of pump connections.

Due to the initial high viscosity of the air at atmospheric pressure, it is necessary to delay the start of the rotors until the backing pump has reduced the pressure to 100 mbar or to use the by-pass valve as shown in figure 3.6(c). This prevents the rotors becoming overheated, expanding and coming into contact with each other or the walls. The initial speed of the rotors can also be controlled by a fluid drive coupling which permits a variable degree of slip during the running up period. Alternatively a suitably controlled by-pass valve can be used during the start-up.

Canned motors are sometimes used as drives for Roots pumps, where the rotor operates in vacuum but the stator windings are at atmospheric pressure, separated by a non-magnetic sleeve. This removes the need for a shaft seal and is often used when clean gas recovery is needed.

3.1.4 Oil free positive displacement pumps

The *diaphragm pump*, figure 3.7, operates by the flexing of an elastomer diaphragm moved by a motor driven piston. They are often two-stage pumps. These pumps will operate to an ultimate pressure of 4 mbar and are used for oil free roughing of UHV systems to allow other pumps to be started (eg adsorption and sputter ion combinations). See chapter 8.

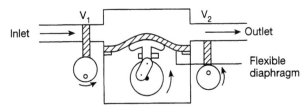

Figure 3.7 A diaphragm pump.

The *multi-stage dry pump** comprises pairs of rotors of different profiles mounted on a common shaft and synchronised by external gearing driven by a directly coupled motor. The first stage rotors form a Roots pump, the succeeding rotors are of claw profile (figures 3.8 and 3.10). The rotors are isolated from the bearings and gear compartments by dynamic seals. Comparative performance is shown in figure 3.9. It is designed for industrial as well as for oil free applications as it can handle condensable vapours and dusty atmospheres.

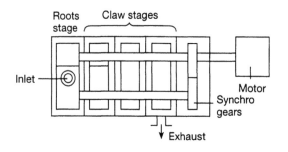

Figure 3.8 Diagram of multi-stage dry pump.

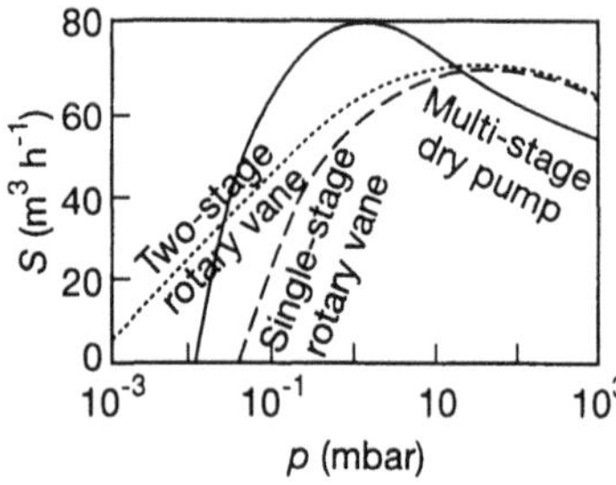

Figure 3.9 Comparative performance of multi-stage dry pump.

* Edwards High Vacuum International.

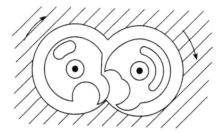

Figure 3.10 Claw profile rotors.

The molecular drag pump described in section 3.2(c) is also a multi-stage dry pump.

The *scroll pump* consists of an orbiting (not rotating) scroll meshing with a fixed scroll. This movement causes pockets of gas to be compressed and passed round to the centre of the pump and discharged to the atmosphere. It has no lubricating or sealing fluids, so it is a dry clean pump. It has an ultimate pressure of 10^{-2} mbar and can be used as a backing pump for turbo- or cryo-pumps. It has been used to provide a stage in the pump-out of a sputter ion-pump.

The *multi-stage piston pump* is a four-stage piston pump, in which each piston is driven by connecting rods and eccentric cams. Drive shafts are permanently lubricated. It has an ultimate pressure of 10^{-2} mbar and is very tolerant of water vapour. Pistons are coated with low friction PTFE while the cylinders are hard coated aluminium and require no lubricants. The pump can be used as a backing pump for turbo- and cryo-pumps. It has been incorporated in a portable leak detector.

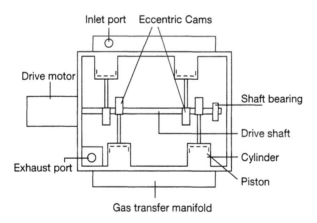

Figure 3.11 Multi-stage piston pump.

Positive displacement pumps for corrosive, toxic and aggressive gases. With the increased demand for the use of hazardous and corrosive chemicals in the semiconductor industry, it has been necessary to develop pumps capable of handling these materials. This involves the use of synthetic fluids which are basically inert to most chemical reactions, and these are mostly from the perfluoro polyether and CTFE groups. Pump fluids are filtered with external circulatory and filtration systems and where toxic or rare gases are being used, it is necessary to ensure leak-tight pump cases and shaft seals. Inert purge gases in the pumps themselves are sometimes used and where flammable gases are being pumped, explosion proofed pumps must be used. Where hazardous gases have to be removed from pump exhausts, hot bed reactors can be used which convert these gases into inert, stable inorganic salts. This conversion takes place in cartridges which, when spent, can be exchanged easily and disposed to normal land-fill sites. Special reactors for semiconductor etching applications are also available.

3.2 KINETIC PUMPS

Kinetic pumps are those in which momentum is imparted to the incoming gas molecules in such a way that the gas is transferred continuously from the inlet to the outlet. Water jet, diffusion, drag and turbo molecular pumps will be described.

3.2.1 The water jet pump

This pump is used in the low vacuum region (lower limit 20 mbar) for example in chemical laboratories for force filtering (figure 3.12). A stream of tap water from a nozzle entrains air or gas in the water jet and carries it to the exhaust thus lowering the pressure at the inlet.

The *vapour jet pump* is a development of the water jet and takes a place between the rotary pump and the diffusion pump having an ultimate pressure between 10^{-2} and 10^{-4} mbar when diffusion pumps become unstable. Their critical backing pressure is high, between 2 and 7 mbar. They are used in the chemical and metallurgical industries when high hydrogen loads can be experienced. Large pumps can have speeds up to $1.5 \times 10^4 \, 1 \, s^{-1}$.

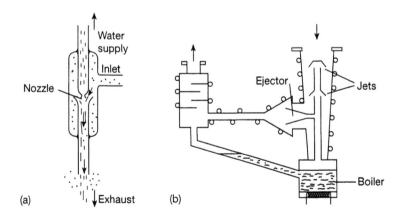

Figure 3.12 (a) A cross-section of a water jet pump; (b) a vapour jet pump.

3.2.2 The diffusion pump

A schematic cross-sectional diagram of the diffusion pump is shown in figure 3.13.

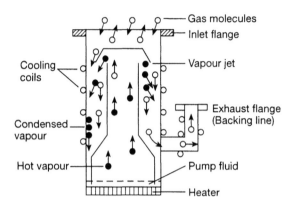

Figure 3.13 Schematic diagram of a cross-section of a diffusion pump.

In this pump a heater vaporises the working fluid which rises inside the vapour chimney and is deflected downwards by the jet assembly in an annular jet at supersonic speed. This high speed jet of fluid molecules imparts a momentum to the random moving incoming gas molecules giving them a direction primarily towards the region of the pump outlet where they are removed by a mechanical pump. The vapour jet condenses on the cooled pump walls and returns to the boiler. This pump operates in the molecular flow region. Pump fluids available are described in section 5.5.

Critical backing pressure

If, during operation, the backing pressure rises to a critical point (typically 0.5 mbar) the vapour jets break down due to the increased gas density in the pump and the high vacuum pressure rises dramatically in the pump and chamber (figure 3.14).

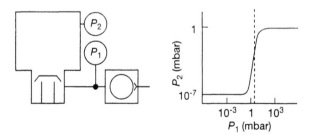

Figure 3.14 The critical backing pressure.

A schematic diagram of a typical diffusion pump vacuum system is shown in figure 3.15 and in chapter 8.

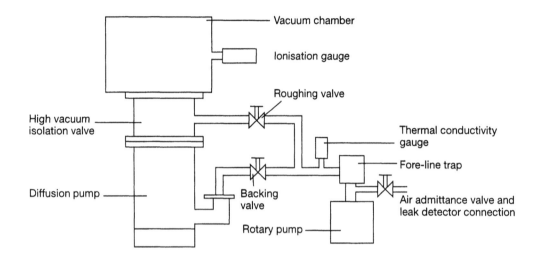

Figure 3.15 A typical diffusion pump system.

An example of a pumping speed curve for a 63 mm inlet flange diffusion pump is shown in figure 3.16. It shows that the pumping speed is more or less constant over a range below 10^{-3} mbar due to the jets being undisturbed by the gas density. The ultimate pressure depends on the vapour pressure of the pump fluid, the pump design and the gas load from the chamber.

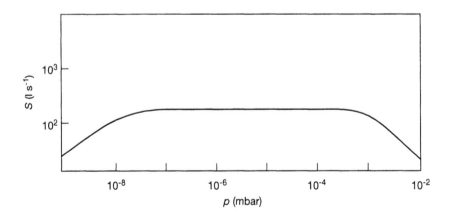

Figure 3.16 Pump speed curve for a 63 mm diffusion pump.

An integrated pump group (figure 3.17) includes in one body without flanges an isolation valve, a cooled baffle and top cap and a diffusion pump.

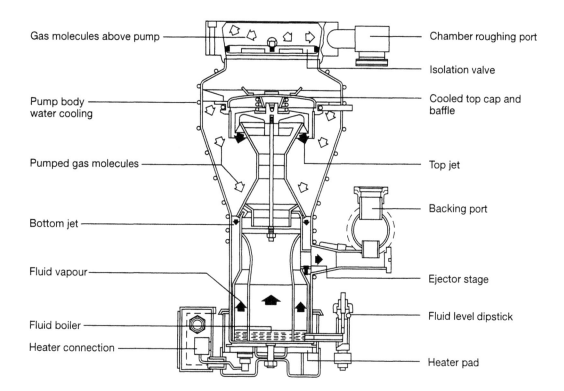

Gas molecules above pump

Chamber roughing port

Isolation valve

Pump body
water cooling

Cooled top cap and
baffle

Pumped gas molecules

Top jet

Backing port

Bottom jet

Fluid vapour

Ejector stage

Fluid boiler

Fluid level dipstick

Heater connection

Heater pad

Figure 3.17   An integrated pump group (Diffstak*). Reproduced by permission of Edwards High Vacuum International, Crawley.

Back-streaming

This occurs when pump fluid molecules move above the upper jet so that they can enter the chamber, causing possible contamination. This can be largely prevented by the use of chilled baffles or a cold trap (as shown in figures 3.18 and 3.19). It can also be greatly reduced by the design of the top jet and by the use of a large cold cap.

The presence of oil vapour in the work chamber is often thought to arise from back-streaming from the diffusion pump, but investigations have shown that this is in fact rotary pump oil which has entered the chamber via the roughing line. When the roughing pressure drops below 10^{-2} mbar rotary pump vapour molecules can move in any direction as they are in molecular flow. This can be greatly reduced by the use of a correctly maintained fore-line trap (figure 3.20). It is important however that if a roughing line is used then once the chamber pressure reaches the correct value, the roughing valve should be closed to minimise the back-streaming of rotary pump vapour into the vacuum chamber.

* Trade name of Edwards High Vacuum International.

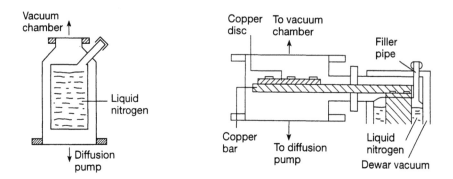

Figure 3.18 A liquid nitrogen cold trap.   **Figure 3.19** A cooled baffle.

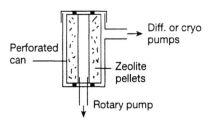

Figure 3.20 A fore-line trap.

3.2.3 The molecular drag pump

The drag pump is a kinetic pump in which a momentum is imparted to the gas molecules by contact between them and the surface of a high speed rotor, causing them to move towards the outlet of the pump. The pump shown in figure 3.21 is part of a multiple pumping system based on the Holweck pump (1923) with the rotor rotating at 27 000 rpm. in a stator with a helical groove. The compressed molecules then enter a dynamic seal (a miniature drag pump) which further compresses the gas to enter a diaphragm pump. The final stage is a dry oil free piston pump or other dry pump.

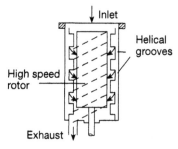

Figure 3.21 A type of molecular drag pump.

3.2.4 The turbo molecular pump

The turbo molecular pump (Becker 1958) is a molecular drag pump in which the rotor is fitted with slotted discs or blades rotating between corresponding discs in the stator, figure 3.22. The linear velocity of a peripheral point of the rotor is of the same order of magnitude as the velocity of the gas molecules. This pump operates in molecular flow conditions.

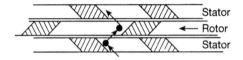

Figure 3.22 Schematic diagram of a rotor and stator discs in a turbo pump.

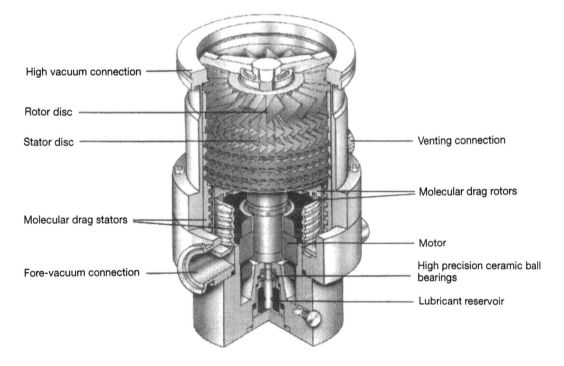

Figures 3.23 A cross section of a turbomolecular/drag pump (Holweck principle), by courtesy of Pfeiffer Vacuum Ltd.

Considerable development has taken place since 1958 when the first pumps were being put into production. Early pumps were dependent on oil or later grease lubricated bearings, but the demand for flexibility in mounting positions and greater cleanliness with freedom from hydrocarbons saw the development and introduction of maintenance-free ceramic ball-bearings instead of steel. These bearings are lubricated for life and have a lower mass, giving lower radial forces reducing running temperature and increased bearing life. Magnetic bearings using high-field permanent magnets have also been successfully developed and now hybrid pumps using a combination of both magnetic and ceramic bearings are commonly available. To improve further the cleanliness of a system it was required to replace the conventional oil-sealed rotary vane backing pump with a 'dry' pump (refer to section 3.1). This requires high compression ratios in the turbo pump so that the backing pressure could be as high as 10 mbar. This is achieved by adding drag stages to the turbo rotor either using the Holweck (1923) principle of spiral channels (figure 3.23) or the Gaede (1913) disc system (figure 3.24). This does not give any increase in size or rotor power consumption.

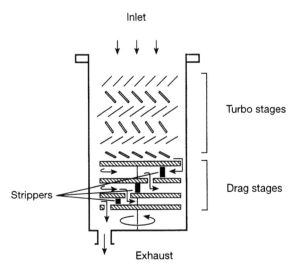

Figure 3.24 Diagram of a turbo molecular/drag pump (Gaede principle), by courtesy of Edwards High Vacuum International.

The early pumps achieved high rotor speeds by a combination of external belts and gearing, later superseded by high frequency motors energised from frequency converters with direct drive to the rotor. Further development has replaced these with brushless d.c. motor drives.

A backing pump is required to exhaust the turbo pump to the atmosphere, as for the diffusion pump, and must be of adequate size for roughing out the pump (and/or the chamber) and maintaining the pressure required for full performance of the turbo. Correct selection is important especially with high input pressures (10^{-1} mbar). Water cooling is generally required to cool bearings and motor especially with high throughput and high inlet pressures.

In the case of a conventional turbo pump without a drag stage, a backing pump is usually a two-stage trapped rotary vane pump. In the case of a compound or wide-range turbo pump with a drag stage, a dry pump of the claw, scroll, diaphragm, or reciprocating piston type can be used, thus giving a totally clean and oil-free system.

An alternative system can be used with a magnetic bearing pump with an added drag stage, exhausting to some form of dry pump when complete freedom from hydrocarbons is required.

When turbo pumps are used with corrosive or abrasive gas mixtures or those having a high O_2 content (25%), a dry nitrogen purge should be used through the purge ports provided. Some pumps are manufactured with nickel coated rotor and stator to supplement the gas purge, and magnetic bearings replace the ceramic ball races.

Where low vibration levels are required (e.g. electron microscopes) the use of all-magnetic bearings is recommended. The use of damped vibration isolating connecting bellows will give further reduction in transmitted vibration.

The turbo pump has a virtually constant pumping speed for all gases between 10^{-2} and 10^{-9} mbar. The compression ratio however is dependent on the gas species, being low for the light gases and a high ratio for the heavy gases and especially hydro carbons.

Table 3.2 Compression ratios for various gases in a turbo pump.

Gas	Compression ratio
H_2	1 000
He	10 000
N_2	1 000 000 000

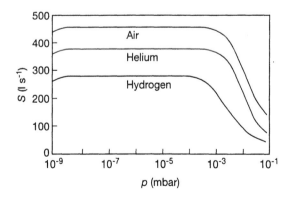

Figure 3.25 Pumping speed curves of a turbo pump for various gases.

It is the very high compression ratio for hydrocarbon molecules that allows the direct connection of the turbo pump to the vacuum chamber without the need for cooled traps and baffles. The same conditions for rough pumping apply to the turbo pumped system as to the diffusion pumped system and the use of a fore-line trap is recommended. Typical pumping speeds for air, helium and hydrogen are shown in figure 3.25.

3.3 ENTRAPMENT PUMPS (CAPTURE PUMPS)

Entrapment pumps are those in which the gas molecules are retained by sorption or condensation on internal surfaces. This implies a limit in capacity and possibly a need for regeneration.

3.3.1 The adsorption pump
This is an entrapment pump in which the gas is retained mainly by physical adsorption of a material of large real surface area (e.g. a porous substance). In the adsorption pump, figure 3.26(a), the zeolite (an alkali alumino-silicate) has a large surface area of 10^3 m^2 per gram of solid substance. One gram of zeolite is capable of adsorbing a monolayer which is the equivalent of 133 mbar l of gas.

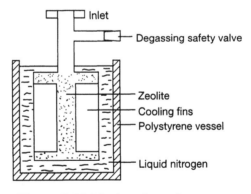

Figure 3.26 (a) An adsorption pump.

When the zeolite is chilled to liquid nitrogen temperature this adsorption process is greatly enhanced and operates until all the adsorption sites are occupied. This pump is used to reduce the chamber pressure for final pumping by sputter ion or cryopumps where complete absence of hydrocarbons is required. Two pumps are often used working alternately as shown in figure 3.26(b) so that when the first pump has taken up its full quantity of gas it can be isolated from the system and the second pump brought into use (chapter 8). The first pump can be regenerated by removing the liquid nitrogen and applying heat, which will drive out most of the adsorbed gases. They have an extremely low pumping speed for hydrogen, helium and neon due to different adsorption properties, but as these gases are only a few parts per million in atmospheric air they present little problem to pumping. In some applications the pressure is reduced initially to 10 mbar by the use of a diaphragm or dry pump then the adsorption pumps are used.

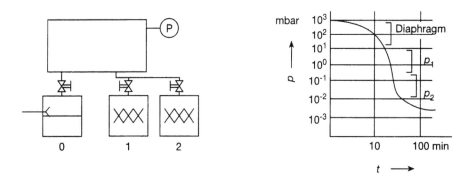

Figure 3.26 (b) Pumps working alternately.

3.3.2 The getter pump

This pump uses the principle of 'gettering' where gas is retained by chemical combination with a 'getter' in the form of metal or metal alloy, either in bulk or in the form of a freshly deposited thin film.

The **sublimation pump** is shown in figure 3.27 and consists of a chamber containing a titanium/molybdenum alloy filament which, when heated, produces a titanium vapour which condenses on any surrounding surfaces. This layer reacts with active gases to form stable compounds (e.g. titanium oxide). The continual production of vapour produces further layers of active titanium. The pumping speed depends on the surface area and temperature of the titanium layer. The pump shown in figure 3.27 has a liquid nitrogen cooled inner surface.

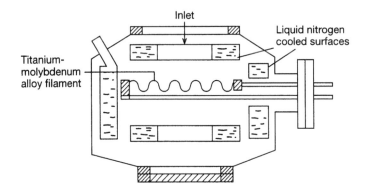

Figure 3.27 A cross-section of a sublimation pump (Pfieffer AG, Liechtenstein).

Table 3.3 The pumping speed for various gases at different temperatures of surface in a sublimation pump.

Temperature of the active getter film	Pumping capacity of new surfaces for various gases at $<10^{-6}$ mbar (l s^{-1}per cm^2)					
	H_2	N_2	O_2	CO	CO_2	H_2O
20 °C	3.0	4.5	9.0	9.0	7.5	3.0
-195 °C	10.0	10.5	10.5	10.5	9.0	14

This pump is used as an additional pump in high and ultra-high vacuum systems to provide high speed pumping during a process, or to reach the working pressure in a shorter time. They are often used for short periods only to conserve the filament material.

The *non-evaporable getter (NEG) pump* consists of a gettering alloy, typically zirconium 84% - aluminium 16% with the trade name of St101 manufactured by SAES Getters SpA, and St707, 70% Zr, 25% Va and 5% Fe. These alloys in powder form are deposited on non-magnetic support strips (e.g. constantan) and arranged to present a large surface area. The material develops its pumping characteristics after an activation process consisting of heating in vacuum. The active gases are pumped by diffusion into the bulk material while the getter is maintained at a high temperature (typically 400 °C) by electrical heating of the backing strip or radiant heater. This pumping process is particularly effective for hydrogen and its isotopes. By heating the getter to higher temperatures (typically 700 °C) the gases can be desorbed. The NEG pump speeds for various gases are given in table 3.4.

Table 3.4 Typical NEG cartridge pumping speeds for various gases.

Pumping speed at 400 °C (1 s^{-1})	H_2	D_2	O_2	CO	O_2
	1 250	900	800	625	210

These pumps have found successful use in plasma machines and distributed pumping in large particle accelerators. They are now incorporated in sputter ion pumps to improve the high pumping speed for hydrogen*. Figure 3.28 shows a pumping element of a large pump where many elements are combined to give high pumping speed. Figure 3.29 shows a pumping element which can be contained in a bolt-on chamber to form a single pump. It is available in two sizes.

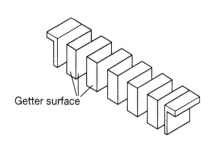

Getter surface

Figure 3.28 An NEG pumping element.

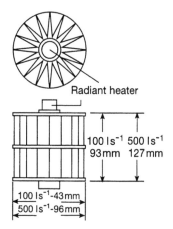

Radiant heater

100 ls⁻¹ 500 ls⁻¹
93 mm   127 mm

100 ls⁻¹-43 mm
500 ls⁻¹-96 mm

Figure 3.29 An NEG pumping capsule.

3.3.3 The sputter ion pump

The sputter ion pump is a getter ion pump in which ionised gas is accelerated towards a getter surface which is continuously renewed by cathodic sputtering. The basic sputter ion pump consists of a flat titanium cathode, a cylindrical anode and an axial magnetic field as shown in figure 3.30.

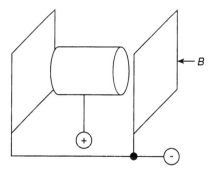

Figure 3.30 Schematic diagram of a one cell sputter ion pump.

In the *diode ion pump* the combined action of the electric (3 -7 kV) and magnetic (0.1 - 0.2 T) fields enables a discharge to be maintained at low pressures. Gas is ionised by energetic free electrons and the positive ions bombard the cathode, sputtering titanium to form getter films on the anode and the opposite cathode (figure 3.31). The titanium reacts with all active gases forming stable compounds,

* Varian, Palo Alto, USA

and also a considerable number of bombarding gas molecules are buried in the cathode.

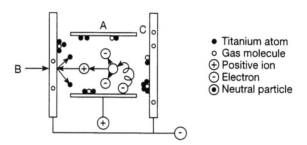

Figure 3.31 A diode ion pump.

Noble gases (He, A, Ne) are pumped by burial under layers of titanium on the pump walls and anode, while others are buried in the cathode. Unfortunately as further sputtering takes place, previously buried molecules can be released, giving rise to instability in pumping. Various solutions to the pumping of noble gases have been attempted, for example the use of differential cathodes where one is titanium and one is tantalum, and the use of slotted cathodes where the bombarding ion arrives at a glancing angle. However the most successful has been the triode ion pump configuration (figure 3.32).

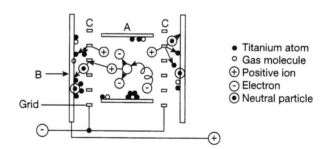

Figure 3.32 The triode ion pump.

In the *triode ion pump* the whole pump body is earthed and, being at the same potential as the anode cylinder, acts as an auxiliary anode. The ions produced as in the diode pump, now graze the titanium grid giving a high sputtering rate, the sputtered titanium forming preferentially on the pump body. Energetic neutrals created by ions glancing off the cathode are buried on the surface of the pump body or are reflected and pumped at the anode. Any positive ions arriving at the pump body are repelled by its positive potential and do not disturb the surface. Buried or implanted noble gases covered with fresh titanium are thus left generally undisturbed, leading to a higher net pumping speed for these gases (table 3.5).

Table 3.5 Relative pumping speeds in a triode sputter ion pump.

Gas	Pumping speed (as % of air speed)
Hydrogen	150 - 200
Methane	100
Oxygen	80
Argon	30
Helium	28

A triode pump with sharp edged cathode cells has been designed (the Varian, 'Star cell' ion pump), which increases the hydrogen pumping speed and raises starting pressure and reduces deformation of the electrodes.

Figure 3.33 shows the pumping speed curve for a typical 150 1 s^{-1} sputter ion pump. It is important to match the power unit to the pump as an overrated power supply will cause overheating and damage to the pump at high pressure and an underrated supply will lower the maximum operating pressure of the pump. The life of a typical diode pump is 40 000 h at 10^{-6} mbar and proportionally longer at lower pressures. It is not a suitable pump where cyclic operations require it to be continually brought to atmospheric pressure.

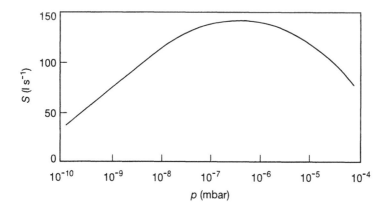

Figure 3.33 Pumping speed curve for a 150 1 s^{-1} sputter ion pump.

Complex molecules such as methane are cracked by electron bombardment and pumped as separate molecules. The pump can be started below 10^{-3} mbar when a discharge can be struck, but with an old pump the power dissipated in the discharge at this pressure can cause heating and subsequent liberation of buried gases. It is therefore advisable to start the pump below 10^{-4} mbar by initial pumping with suitable roughing pumps (section 3.3.1). The pump current is proportional to pressure so that the current meter on the power unit has a range calibrated in millibar. The sputtering rate is also proportional to pressure, so the pump speed is reasonably constant.

3.3.4 Cryopumps

In a cryopump the gases are condensed on a cold surface and retained there. It is a clean pump, free from hydrocarbons and its pumping speed is proportional to the surface area refrigerated. Figure 3.34 shows the temperatures required to maintain an equilibrium vapour pressure equal to or less than the required low pressure in the chamber.

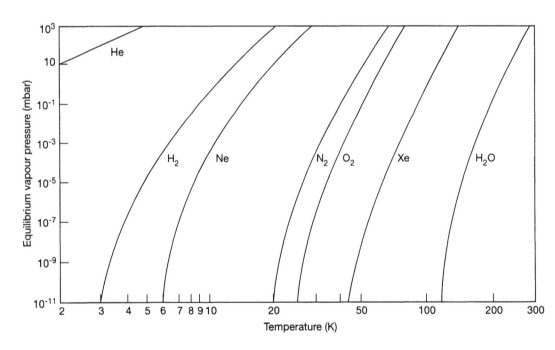

Figure 3.34 Temperature/vapour pressure for various gases.

There are two main types of cryopump.

In the *liquid pool cryopump* the pumping surface temperature is maintained by liquid helium, thermally shielded by a jacket and baffle cooled with liquid nitrogen as shown in figure 3.35.

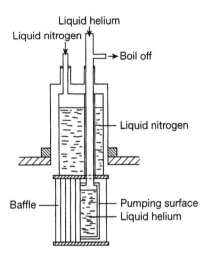

Figure 3.35 A liquid pool cryopump.

This pump requires large amounts of liquid helium but can have high pumping speeds (eg 107 l s⁻¹) with specially designed surfaces and baffles.

The present day bolt-on ***self-contained cryopump*** is cooled by an internal two-stage refrigerator supplied by a room temperature external helium compressor. The pump is contained in its own vessel and a cross-sectional diagram of such a pump is shown in figure 3.36.

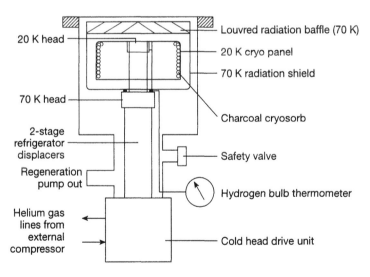

Figure 3.36 A diagram showing a cross-section of a self-contained cryopump.

As can be seen in figure 3.34 the 20 K surface can reduce the vapour pressure of all gases except hydrogen, helium and neon to below 10^{-11} mbar. The latter gases are pumped by cryosorption in cooled charcoal granules glued to the underside of the

20 K element, this position protecting it from other gases. Water vapour is pumped on the 70 K baffle and radiation shield. The cryopump operates in the pressure range below 10^{-2} mbar to reduce thermal loading on the cryo-surfaces by conduction. The cryo-panels are cooled by a cryo-cooler using Gifford - McMahon or modified Solvay cycle refrigerators where helium from the compressor enters at room temperature at about 13 bar pressure. Two stages of compression and expansion produce the 70 and 20 K temperatures at the heads and the compressor provides filtering and cooling of the circulating helium. Maintenance of the adsorber in the compressor is required, about every 12 -15 000 hours.

The cryopump has to be regenerated at intervals depending upon the gas load pumped, and when the cryo surfaces are fully 'loaded' with condensate. With the pump and compressor switched off, the pump is allowed to warm up, often aided by the use of a dry nitrogen purge. Before further use it is pumped out to the required starting pressure. Care must be taken during the regeneration process, as gases liberated individually as the temperature rises can quickly produce high pressures within the pump requiring the use of a self-sealing safety valve and perhaps ducted exhaust when explosive or aggressive gases have been pumped. The cool-down time is typically one hour and pumps are available with pumping speeds up to 10 000 l s^{-1}. Some displacers have a speed control, which in effect controls the cooling power, giving stand-by operation, and boost performance, to economise in piston wear and helium consumption. The cryopump is chiefly used when a hydrocarbon free system is essential, as in surface science applications, semiconductor production, particle accelerators and space simulation chambers. Figure 3.37 shows a typical cryopumped system. NB. One He compressor can supply a number of cryo pumps.

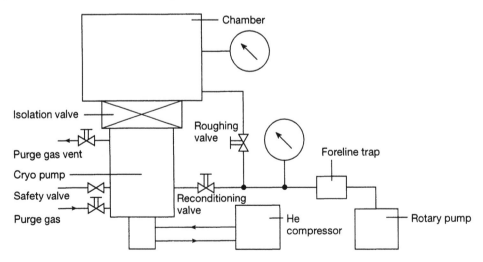

Figure 3.37 A typical cryopumped system.

3.4 PUMP SELECTION

The correct selection of a pumping system will depend on many factors. A number of questions should be asked in order to select the appropriate pump. These are given below and should be used in conjunction with the working ranges of pumps as shown in table 3.6 and range required for various processes, table 3.7.

How to select the right pump.

1   What do I want to do? What is the process?
2   Ultimate pressure required?
3   Total volume and surface area of the chamber and the surface outgassing rate?
4   Operating pressure required (including the input gas load)?
5   Maximum tolerable pressure?

6   Pump-down time required from atmosphere with all pumps conditioned (cycle time)?
7   Anticipated working life of system?
8   Services available, supplies of liquid gases, any special venting or gas recovery requirements?
9   Dimensions and weight (including headroom and lifting equipment)?
10   Skills of operators available?

11   Special requirements?

(a) Ultra-clean
(b) Hydrocarbon free
(c) Toxic or corrosive gases
(d) Reactive gases
(e) Process temperature
(f) Bakeout
(g) Radiation environment
(h) Rare gas recovery
(i) Exhaust discharge

12   Vibration limits?
13   Economics - liquid nitrogen, power, fluids, capital available?
14   SAFETY!!

Table 3.6 The working pressure ranges of vacuum pumps.

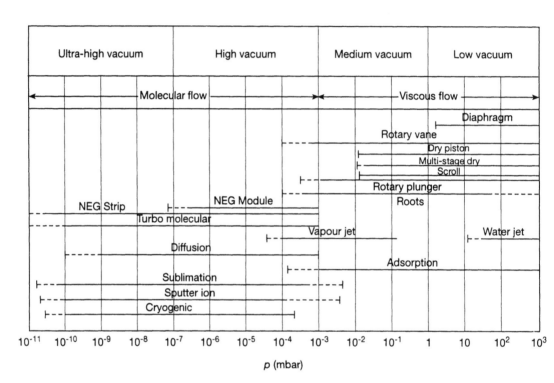

Table 3.7 Working pressure ranges for industrial and research processes.

Industrial	UHV	High	Med	Low
Annealing of metals				
Metal melting				
Molecular stills				
Evaporation coating				
Metal de-gassing (casting)				
Electron tube production				
Freeze drying				
Electron beam welding				
Scientific research				
Mass spectrometers				
Electron microscopes				
Thin film production				
Surface science				
Electron diffraction				
Particle accelerators				
Storage rings				

10^{-13} 10^{-7} 10^{-3} 10^{0} 10^{3}

p (mbar)

Summary
This chapter has described examples of pumps in common use, chosen to show principles of operation in the various groups. It is realised that it is by no means exhaustive and does not cover innumerable variations and manufacturers special features.

For further information, the special edition of *Vacuum* entitled 'Modern vacuum practice' Vol 37 Nos 8/9 1987 edited by G F Weston is recommended. Manufacturers' literature is also recommended.

Reference

Becker W 1958 *Vakuum-Technik* **7** 149

4

Measurement of Pressure

R K Fitch and A Chambers

The measurement of the total pressure in a vacuum system is just as important as the production of the vacuum. However because of the extremely large range of pressures involved - from atmospheric down to 10^{-12} mbar or less - there is no single vacuum gauge which covers the whole range, which is perhaps not surprising. Choosing the gauges for a particular application is therefore an important consideration and the correct interpretation of their readings is essential. For example, in many cases it may be satisfactory to note that a gauge reading is about the same as it was on the previous day, but in some situations it may be necessary to record and interpret quite small changes in pressure over the same period. Thus concerns about stability of a gauge and the accuracy of its calibration arise. Furthermore for many types of gauge the reading depends on the composition of the residual gas and so this may have to be taken into account in evaluating the indicated pressure. It may be that only the total pressure is of interest but in many applications of vacuum technology a knowledge of the composition of the residual gas is also important and therefore it is necessary to incorporate a partial pressure analyser in the vacuum system. Thus it is convenient to divide the discussion into 'total pressure' and 'partial pressure' gauges.

4.1 TOTAL PRESSURE GAUGES
Since the earlier part of this century a large variety of vacuum gauges has been developed but only a relatively small number of types have been commercially successful and remain in common use. The main emphasis will be on these devices, although the details and significance of a small number of important specialised devices will also be given. The operation of total pressure gauges may be classified as depending on one of three different physical properties, namely hydrostatic pressure, thermal conductivity and electrical ionisation.

4.2 HYDROSTATIC PRESSURE GAUGES
One of the oldest types of vacuum gauge, the *liquid level manometer*, usually consists of a simple U-tube manometer in which one arm of the tube is either open to atmospheric pressure or closed and permanently evacuated to a pressure much

lower than the pressure to be measured, so that the difference between this reference level and the liquid level in contact with the vacuum is measured directly. The gauges are absolute devices and with a low density vacuum pump fluid of low vapour pressure they can be used from atmospheric pressure to about 1 mbar. In the appropriate circumstances their simplicity and accuracy make them very attractive devices and, if sophisticated optical techniques are used to measure the height of the column, then they can be used down to about 10^{-5} mbar, as discussed by Berman (1985). A familiar hydrostatic device is the **McLeod gauge** which until the early 1980s was used as a standard for calibration purposes. It is now obsolescent but its principle of operation remains fundamental and of interest. A simplified version of it is shown schematically in figure 4.1.

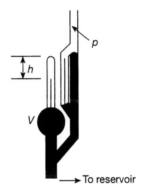

Figure 4.1 Schematic diagram of the McLeod gauge.

The gas is sampled from the vacuum system at the prevailing pressure as the mercury reservoir is raised to isolate a sample of volume V at pressure p. Thus if the cross-sectional area of the capillary is A mm^2 and the height of the column is h, then by Boyle's law, with pressure expressed in mm of mercury and h measured in mm,

$$pV = (h+p)Ah$$

and $p << h$ so that

$$p = Ah^2/V$$

The McLeod gauge is normally only useful down to a pressure of about 10^{-5} mbar. The pressure scale is non-linear and continuous readings cannot be taken. Because it is a compression device problems can arise with condensible vapours and so the use of a chilled trap may be necessary condense them out. A simple and still useful version of the McLeod gauge is the 'Vacustat' which covers the range 1 to 10^{-3} mbar, is easily portable, and can be used to check other gauges. However the use of toxic mercury makes these gauges less acceptable under the present health and safety regulations.

The *capsule gauge* is a simple and robust mechanical device which measures pressures from 1000 down to 1 mbar by sensing and magnifying mechanically the small changes in dimension of a capsule which expands as the pressure surrounding it is reduced. The capsule consists of two thin walled corrugated diaphragms welded together at their edges and hermetically sealed, and is located inside a leak-tight housing which attaches to the vacuum system. An internal lever mechanism moves the pointer over a scale within the housing. Because of their construction they are not affected by changes in atmospheric pressure and are approximately linear in response. They are supplied by a number of manufacturers and available for ranges such as 0 - 25 and 0 - 100 as well as 0 - 1000 mbar with accuracies quoted as a few percent of their full scale deflection. They are rugged and immediate indicators of the vacuum they measure, although the possibility of the mechanism becoming impaired due to corrosion by hostile gases for example or other reasons may be a concern.

In *diaphragm gauges* the pressure difference across a single circular corrugated diaphragm causes a deflection which may be coupled with mechanical magnif-ication to a pointer and scale. An ingenious version of the device described by Wutz *et al* (1989) and manufactered by the Leybold Company operates over the range 1000 - 1 mbar with good accuracy and an expanded pressure scale in the lower range so that 1 - 20, 20 - 100 and 100 - 1000 mbar occupy comparable lengths of a generous 280° scale. This is accomplished by having a large very sensitive diaphragm and a set of concentric circular stops which render it progressively less sensitive at higher pressures.

The *capacitance manometer* is another type of gauge exploiting the pressure dependent deflection of a diaphragm. In recent years it has come to be regarded as a most reliable and accurate total pressure gauge and is widely used. A schematic diagram of the sensor of such a device is shown in figure 4.2 in which two capacitance electrodes, one in the form of a circular disc (D) and the other in the form of a circular annulus (A), are deposited on a ceramic substrate (S). This is positioned close to an inconel diaphragm which forms two capacitors - one with the disc and the other with the annulus - and which form part of an AC bridge circuit.

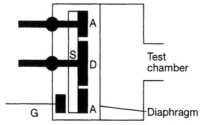

Figure 4.2 Capacitance manometer.

In the 'absolute' (as distinct from the differential) form of the manometer the inconel diaphragm is exposed to the gas under test; the other side is sealed off under a vacuum of less than 10^{-7} mbar, after which a chemical getter (G) is activated to maintain the low pressure. When the pressures on each side of the diaphragm are equal, the bridge can be balanced. As the pressure in the test chamber is increased the diaphragm deflects towards the opposite electrodes, the central capacity changing more than the outer capacity, thus causing the bridge to become unbalanced. The output of the AC bridge is amplified and demodulated to produce a high level DC output signal of order 10 V. The highest pressure measurable is when the diaphragm touches the electrode which also acts as an over-pressure stop. These manometers have now reached an advanced stage of development and were shown by Sullivan (1985) to be very accurate, sensitive, reproducible and reliable, and operable up to temperatures as high as 300 °C. However one of their main advantages is that being total pressure gauges their reading is independent of the gas composition and they can therefore be used to monitor reacting gases, as is necessary in chemical vapour deposition (CVD). They are also essential in semiconductor processing technology when the gas pressure has to be accurately servo-controlled from the gauge reading for wafer production and in growing single crystals of silicon when the gauge must be insensitive to RF induction.

Each sensor will normally operate over four decades and thus only a small number of sensors is required for use from atmospheric pressure to their present limit of 10^{-5} mbar. The main disadvantage of these gauges is that they are very sensitive to temperature changes and in order to achieve their optimum performance, accurate temperature control is essential. They were shown by Nash and Thompson (1983) to be very reliable as a transfer standard. Manufacturers give uncertainties of 0.5% at 1 mbar and only 1% at 10^{-3} mbar. These gauges have always been considered to be quite expensive though in fact the actual cost has not increased much in the last decade.

In the differential form of the device the pressures on both sides of the sensor diaphragm are variable so that their difference can be detected with a sensitivity and range which depends on the particular choice of sensor selected. This can be exploited for example in situations in which one wishes to control the pressures of different gases in feeder lines on opposite sides of the sensor, prior to their being mixed downstream.

4.3 THERMAL CONDUCTIVITY GAUGES

In these gauges the pressure dependent conduction of heat through a gas from a hot resistively heated fine wire filament to a surrounding glass or metal envelope at room temperature is the basis of the pressure indication. In the *Pirani gauge* current is driven through the filament by incorporating it into a bridge circuit as shown in figure 4.3, in which a driving source of voltage V sends current through both the filament branch ($R_F + R_1$) and the branch ($R_C + R_2$). A detector D which can be a simple meter detects out-of-balance current when $R_F/R_1 \neq R_C/R_2$. The filament is usually made of tungsten, which has a good temperature coefficient of resistance. To allow for changes in ambient temperature the resistor labelled R_C is sometimes a 'dummy' filament isolated and encapsulated at high vacuum alongside the sensing resistor or more frequently nowadays a temperature compensating arrangement.

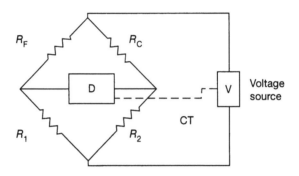

Figure 4.3 Circuit of the Pirani gauge.

Although as noted in section 1.13 the thermal conductivity of gases at normal pressures is constant, below about 100 mbar the heat conducted between filament and envelope starts to fall, eventually becoming proportional to the gas density and hence pressure. This is because the gas ceases to be in a continuum state, so that heat transfer increasingly involves direct flight of 'hot' molecules from the filament surface to surfaces at room temperature as pressure falls and mean free paths become comparable with and exceed the gauge dimensions. For a range of pressure from about 5 down to 1×10^{-3} mbar, heat loss from the filament is dominantly by conduction through the surrounding gas; at the lower value it has become comparable with the loss due to radiation which is small at the temperatures used, typically about 100 °C. There is also a small heat loss through the filament supports at its ends. At the upper end of its pressure range, above about 10 mbar, in addition to the constant conduction loss through the gas, there is a small loss due to convection which is proportional to pressure.

The simplest and traditional way of operating the gauge is as a *constant voltage* device. In this case the driving voltage V is held at a constant value and as the pressure alters so does the cooling rate of the filament, causing a change in its temperature with a consequent change in its electrical resistance, and hence a change of the out-of-balance current in the meter. With decrease of pressure resistance increases. On analogue instruments the range of 10 to 1×10^{-3} mbar is displayed on a generous 230° scale which becomes 'squashed' at both ends due to the non-linear overall response, greatest sensitivity being between 10^{-2} and 1 mbar. The instrument may be adjusted by the user at the upper or lower ends of the scale by balancing the bridge at atmospheric pressure or high vacuum, 10^{-4} mbar or less, respectively.

In another and more expensive form of the instrument operation is in the *constant temperature* mode. In this case modern solid-state electronics are used to hold the bridge in balance by regulating the driving voltage V by the detector, as suggested by the representative dotted link CT in figure 4.4. The variation of V controls the power I^2R_F supplied to maintain the filament temperature and hence resistance constant. Greater cooling at higher pressures requires larger I values to maintain bridge balance. The pressure value displayed is derived electronically from the power provided. Because the temperature is constant the instrument's response to pressure changes is rapid and the electronic control allows the range to be extended up to several hundred mbar. Leck (1989) and Wutz et al (1989) describe the theory of the instrument and its operational modes in detail.

The *thermocouple gauge* operates on the same filament-cooling principles as the Pirani gauge but in this case the changes in temperature of the filament due to pressure changes are measured directly by a very fine thermocouple attached to its centre. The thermocouple voltage is processed to give a pressure reading. Its working range is typically from 5 down to 1×10^{-3} mbar.

Both Pirani and thermocouple gauges record the total pressure for gases and vapours and because the thermal conductivities of the various gases encountered can differ by as much as twenty times, great care should be taken when interpreting the gauge readings. Their reproducibility at best is about ±10%. The complexities of the heat transfer processes at surfaces and their dependence on the state of the filament are such that pressures cannot be reliably predicted by analysis of the primary voltage/current readings and so calibrations are necessary. The gauges are usually calibrated for nitrogen or dry air, and manufacturers supply response data for other common gases. In an excellent text Bigelow (1994) discusses in detail the practicalities of using thermal conductivity gauges.

There is no significant benefit in extending the low pressure range of these gauges but there is a demand for them to operate up to atmospheric pressure. Durakiewicz and Halas (1995) describe a novel and inexpensive electronic way of enhancing the operation of a commercial Pirani gauge using a special bridge circuit so that it measures up to 100 mbar. A number of manufacturers have exploited the small convection loss referred to earlier to extend the range of the Pirani type devices towards atmospheric pressure so that the range they span is 1×10^{-3} to 1000 mbar. This involves modification of the head design to promote convective losses, and sophisticated electronic processing.

The thermal conductivity gauges are likely to continue to play an important role in a majority of our vacuum systems to monitor, for example, the performance of the rotary or sorption pump in primary pumping from atmospheric pressure, and in continuous monitoring of backing line pressure for secondary high vacuum pumps. The rapid response in the constant temperature mode makes them particularly appropriate as sensors for control applications. Though both types can survive sudden exposure to atmosphere, the thermocouple gauge is often preferred because it is more robust.

4.4 IONISATION GAUGES

Ionisation gauges cover the pressure range 10^{-3} to 10^{-10} mbar and less and exist in two main forms namely the 'thermionic ionisation gauge' and the 'cold cathode gauge'.

4.4.1 Thermionic ionisation gauges

These gauges rely on the fact that an energetic beam of electrons (e^-) will ionise the molecules (A) in a low pressure gas according to the equation

$$A + e^- = A^+ + 2e^-$$

while the resulting ion current, I^+ is related to the electron current, I^-, by the equation

$$I^+ = nl\sigma I^-$$

where n is the molecular density, l is the average electron path length and σ is the 'ionisation cross-section' or probability of ionisation.

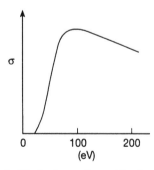

Figure 4.4 Variation of σ with electron energy.

Figure 4.4 shows how σ varies with electron energy; to give maximum ionisation the electron energy is fixed at about 150 eV. The cross-section varies for different gases so the gauge is normally calibrated using dry nitrogen. At low pressures, σ

and l are constant and hence the sensitivity, K, for the gauge at a pressure p is defined by the equation

$$K = (I^+ / I^-) / p \quad \text{mbar}^{-1}$$

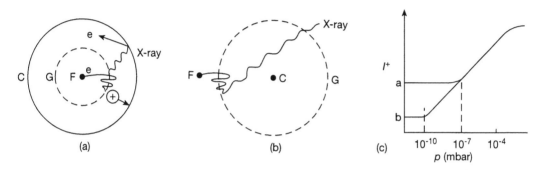

Figure 4.5 Schematic diagrams of (a) conventional ionisation gauge, (b) BA gauge and (c) variation of ion current with pressure for both gauges.

A schematic diagram of a conventional gauge is given in figure 4.5(a) which shows the thermionic filament, F, the cylindrical open mesh grid, A, and the ion collector, C. The collector is at earth, the filament at +30 V and the grid at +180 V potential. The electrons emitted from F are accelerated to the grid where some strike it but a majority pass through to enter a region of decelerating electric field and oscillate back and forth through the grid before being eventually collected there. Some of these electrons will ionise the gas molecules and the positive ions produced are attracted to the earthed collector, C, to produce the ion current. The value of K is in the region 10 - 20 mbar^{-1} and the gauge has a useful range of about 10^{-3} - 10^{-7} mbar. The upper pressure limit, shown in figure 4.5(c), is due to (1) space charge effects, (2) ionisation by the secondary electrons and (3) excitation rather than ionisation by the electrons as the pressure increases. The more important low pressure limit arises mainly from a residual current of secondary electrons, which are released from the collector by soft x-rays generated by electron bombardment of the grid A. Electrons leaving the collector are indistinguishable from positive ions arriving. The effect is independent of the pressure, see figure 4.5(a).

Further contributions to the residual current include electron desorption of adsorbed molecules - particularly oxygen - from the grid as positive ions, referred to as either the EID (electron impact) or ESD (electron stimulated) desorption contribution to the collector current. In another effect - the reverse x-ray effect - many of the x-ray photons miss the collector and impinge on the gauge envelope releasing electrons, of which some can reach the collector. The low pressure limit can be reduced by using the electrode geometry as shown in figure 4.5(b) where a fine wire collector is at the centre of the grid and the filament is outside it. With this arrangement the interception of the x-rays by the collector is reduced by about 100 times and the gauges can be used down to a pressure of about 10^{-10} mbar. This is known as the *Bayard - Alpert gauge* (BAG). It played an important role in the development of ultra-high vacuum techniques in the 1950s. In fact it was fortunate that the BAG was very inefficient at collecting the electron desorbed ions because they have sufficient energy and angular momentum to orbit round the collector and escape through the grid, so that only about 1% of these ions reach the collector. See Leck (1970).

The advantages of the thermionic ionisation gauge are that it is a very reliable gauge and straightforward to operate. It can easily be de-gassed by electron bombardment if the electron current and grid voltage are increased to provide about 35 W of heating power. Such degassing helps to reduce the ESD current referred to above. These gauges have a linear calibration of ion current with pressure but the sensitivity of the BAG tends to drift more than that of the conventional gauge.

A disadvantage of these gauges is that they incorporate a hot filament that can 'burn out' due to accidental exposure to atmospheric air and so most are fitted with two switchable filaments. They also have significant ionic and electrical pumping effects which produce a lower pressure in the gauge envelope than in the vacuum chamber, so that in most UHV systems the gauge is used 'nude' to reduce these effects, meaning that it is exposed directly to the system rather than being housed in a communicating envelope. Furthermore chemical and thermal reactions take place with the tungsten filament at 2700 K which result in the production of undesirable gases such as carbon monoxide and methane. Hence in many applications it is advantageous to use a low work function emitter such as lanthanum hexaboride or rhenium which will give the same emission at only 1500 K. Even clean rhenium is to be preferred to tungsten at the same temperature because it is less chemically active.

Another version of the BAG is the *modulator gauge*, MBAG, in which an additional electrode is added to modulate the ion current from the gas phase ionisation. Its purpose is to enable the normal and x-ray induced components of the collector current to be distinguished. The modulator electrode is within the grid structure and can be at either grid potential V_1, when it scarcely affects the ion current, or at ground potential $V_2 = 0$, when the sensitivity to ion currents is markedly reduced by the modification caused to the electric field. The currents I_1 and I_2 are correspondingly given by

$$I_1 = I^+ + I_R$$

and

$$I_2 = (1 - k)I^+ + I_R$$

where I_R is the residual current due to x-rays and k is assumed to be a constant. Thus we have

$$k = (I_1 - I_2) / I^+$$

and

$$I_R = I_1 - (I_1 - I_2)/k$$

The constant k, which is typically about 0.6, can be determined by operating at a higher pressure, when I_R is negligible compared with I^+. In low pressure operation knowledge of I_R then allows a corrected value of the pressure to be deduced. Many applications of the modulating technique have been reported, an example being its application in the CRYRING heavy ion accelerator/storage system which operates down to pressures of less than 10^{-12} mbar (Lindbland et al 1987). However there are concerns with the use of modulator techniques in general that the x-ray and ion desorption effects are not completely independent of the modulator potential, so that there is likely to be a large uncertainty at these very low pressures.

It is very important to realise that the sensitivity of ionisation gauges varies for different residual gases because of their different ionisation cross-sections and therefore ionisation gauges are normally calibrated using dry nitrogen. When the gauges are used with other gases the true pressure can be calculated from the measured pressure using the equation

$$p \text{ (true)} = p \text{ (measured)} / K$$

in which typical values of K for various gases are:

Nitrogen	Dry air	Hydrogen	Helium	Argon	Neon	Carbon monoxide
1	1	0.4	0.18	1.4	0.34	1.6

K values less than unity such as that for helium reflect the fact that helium is more difficult to ionise than nitrogen, so that the true pressure is considerably higher than the one indicated. However even when this is taken into account care should be taken in interpreting the gas pressure because it is known that gauge sensitivities can vary from one manufacturer to another by as much as ±40%. See Nash (1987). More recently Arnold and Borichevsky (1994) have made a comprehensive study of the stability and accuracy of a large range of hot and cold cathode gauges over extended periods, to which Tilford et al (1995) have added some comments. For most, changes of sensitivity of several tens of percent are observed.

4.4.2 Cold cathode ionisation gauges

An operational problem which arises with the thermionic ionisation gauge is the occasional failure of the hot filament due to an accidental mishap. This is avoided in the *Penning gauge*, which utilises a cold cathode discharge.

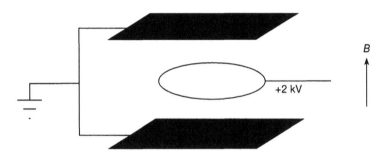

Figure 4.6 Schematic diagram of Penning gauge.

A schematic diagram of the gauge is given in figure 4.6 which shows a ring anode between two flat cathode plates, with an axial magnetic field of about 0.05 T and an anode voltage of 2 kV, the magnetic field normally being provided by a permanent magnet. More recent designs may have a short cylindrical anode rather than a ring. The combined crossed electric and magnetic fields causes any electron present to achieve very long path lengths before collection at the anode, thus increasing the probability of ionisation and enabling the discharge to be self-maintained at low pressures. The discharge is entirely electron-space-charge controlled because the ions are extracted rapidly from the volume and thus once the discharge has stabilised, the pressure p is related to the discharge current i by the equation

$$i = k p^n$$

where k and n are constants in which the index n has a value in the range 1.1 to 1.2. The main advantages of this gauge are that it is very robust, it has no thermionic filament, no x-ray limit and does not produce any thermal radiation. However it is normally considered to be less accurate than the thermionic gauge because i is not directly proportional to p. It also has a relatively large pumping speed, ~0.1 l s^{-1}, and sometimes there are problems with striking the discharge at low pressures. Nevertheless it is easy to operate and is widely used for many scientific and industrial processes in the pressure range 10^{-3} to 10^{-7} mbar when high accuracy is not essential. See Redhead (1987) for a full discussion of its characteristics.

Another type of cold cathode gauge, the *inverted magnetron* of Hobson and Redhead (1958), operates down to much lower pressures in the UHV range. The construction of a typical device is shown schematically in figure 4.7(a). A strong external magnet produces an axial magnetic field in a cylindrical electrode geometry in which the outer cylinder is a cathode and a central axial rod forms an anode A at a potential of several kV. An auxiliary electrode G at cathode potential diverts field emission currents to itself rather than cathode C which collects the ion current. The electric field is radial and perpendicular to the magnetic field. As in the Penning gauge the crossed fields lengthen electron paths which take the form of cycloidal hops in a general circumferential direction as shown in figure 4.7(b). A stable discharge builds up in which the current depends on pressure. The ions produced, being much heavier than electrons, follow simpler trajectories to the cathode. The current - pressure relationship is again $i = kp^n$ with n in the range 1.05 to 1.25 depending on the particular design; control units are designed to translate this current into a displayed pressure value. Typically k has a value of a few A mbar^{-1}at 10^{-6} mbar, taking n as unity at this pressure, and gauges may be used down to 10^{-10} mbar or less, when currents become very small and concerns about spurious leakage currents arise. The time for the discharge to become established can become rather long at lower pressures (typically 30 seconds at 10^{-9} mbar and two minutes at 2×10^{-10} mbar) and so if circumstances allow the gauges are started at high pressure.

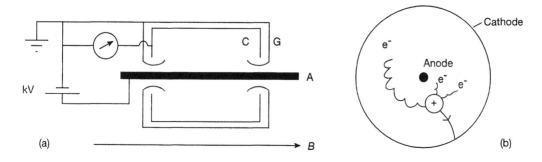

Figure 4.7   Inverted magnetron gauge: (a) construction; (b) discharge geometry.

As with the Penning gauge there are pumping effects of order 0.1 l s^{-1}. They may be noticed, particularly in small volumes, by the changes of pressure which occur on a partner gauge, if one is available, when the IMG is switched off. Operation for prolonged periods at higher pressures $\sim 10^{-5}$ mbar or higher is to be discouraged because of substantial sputtering of electrode materials onto insulators leading to spurious leakage currents and hence false readings. Also insulating surface coatings which may be deposited onto conducting electrode surfaces due to breakdown of hydrocarbons can alter the electrical conditions of the discharge and hence the calibration. Because the power consumed in normal operation is so small outgassing is minimal. Any x-ray effects are proportional to current and scale with pressure, and so do not present a problem. Because it has no hot filament to influence the gas composition by promoting chemical reaction the IMG arguably intrudes less on the vacuum measurement than does the thermionic gauge.

The relative merits of hot and cold cathode gauges have been examined in detail by Peacock *et al* (1991). More recently Kendall and Drubetsky (1997) have evaluated the performance of a number of cold cathode gauges for UHV work and make the case that they offer comparable operating accuracies to those available using well operated thermionic gauges. The cold cathode gauges in various forms continue to be a very useful devices in high and ultra-high vacuum. They have a number of very special applications such as when, for example, the visible radiation from a hot filament cannot be tolerated

4.5 SPECIAL GAUGES FOR THE LOW UHV AND XHV REGIONS

Since the early days of ultra-high vacuum there has been a strong interest in developing ionisation gauges which could operate down to extremely low pressures. The book by Weston (1985) contains a good introduction to the problems involved and the range of special devices invented; that of Leck (1989), on total and partial pressure measurement gives a thorough and comprehensive discussion of principles and practice. The classic text of Redhead *et al* (1968), recently reissued, presents an account of developments to its date, many still relevant and significant. In more recent reviews (1987) and (1993), Redhead has discussed progress in the subject, which remains an active field of development.

The two principal problems which face the designer of gauges to operate in the region below 10^{-11} mbar are caused by the contributions to the collector current of (1) electrons released by x-ray impingement and (2) ions desorbed by electron impact. To a lesser extent outgassing from hot surfaces is also a concern. Although modulation techniques as previously described enable the magnitude of the x-ray effect in the slightly modified Bayard - Alpert geometry to be estimated, allowing pressure estimates down to $\sim 10^{-12}$ mbar, uncertainties increase and other gauges have been developed to try to eliminate it. They are known as *extractor gauges* because they are designed to extract the ions generated in the gas phase to a remote collecting electrode which is shielded from the x-rays produced at the grid. The basic Helmer and Hayward (1966) design which has come to be known for obvious reasons as the 'bent-beam' gauge is shown in figure 4.8(a). The ions are extracted from the ionising volume inside the grid and are electrostatically deflected through 90° onto the ion collector (C). A suppressor grid (S) placed in front of the collector reduces any secondary or photoelectric emission from the collector. The electron and ion ballistics are rather complicated. Its low pressure limit is about 10^{-12} mbar. It has been further developed in a number of ways which are reviewed by Redhead

(1993). Thus Benvenuti and Hauer (1980) described an improved version of the gauge which could be used to 10^{-13} mbar. These gauges also give some reduction of ESD effects because in the process of ion extraction the energy filtering properties associated with the beam-bending electric fields can favour the passage of the 'proper' gas phase ions from the centre of the ionisation volume. Some of these special gauges have x-ray limits as low as 10^{-14} mbar.

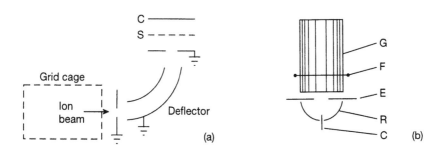

Figure 4.8 (a) Helmer pattern; (b) Leybold extractor gauge.

A pattern of extractor gauge due to Beeck and Reich (1972) and manufactured by the Leybold Company has been available commercially for some time and is frequently quoted as a reliable gauge of reference by workers developing other devices in this low UHV region. Its principle is shown in figure 4.8(b). A ring filament F at +100 V potential surrounds an open grid structure G at +220 V. Shield E at earth potential hides the small collector C from grid generated x-rays because of its small aperture and at the same time positive ions are attracted to it and through the aperture into the collection region. The small collector wire C is surrounded by a reflector R at +205 V potential which intercepts most ESD ions from the grid whilst those formed within the grid are directed to C because they originate from a region of somewhat different potential than the grid itself, due to the effects of the space charge. Measurements down to 10^{-12} mbar are possible. Watanabe (1996) observed a sensitivity reduction of up to 30% in this type of gauge due to accumulated collector contamination with use, and presented a simple means of assessing and curing it to restore sensitivity.

The field of XHV total pressure measurements is in active state of development. For example Watanabe (1993) reported a sophisticated design of *ion spectroscopy gauge* incorporating a hemispherical energy analyser which enables separation of the ESD and gas phase ions. The x-ray limit of this gauge is reported as 2.5×10^{-13} mbar and by an optimal choice of materials and heat conduction paths temperature rises are restricted so that outgassing is reduced to a very low level.

In a new type of hot cathode ionisation gauge developed by Akimachi *et al* (1995, 1996, 1997) and called the *axial symmetric transmission gauge* the ion source and ion detector, which is a secondary electron multiplier are separated by an energy analyser of the Bessel-box type. The design prevents direct axial line of sight from ion source region to detector as well as acting as an energy analyser. Gas phase ions are separated from ESD ions because the energy filter collects the former from a restricted volume of the grid cage where the potential is about 30 V different from the grid voltage. Calibration involves an especially developed conductance modulation method. The sensitivity is 0.23 per mbar for H_2 from 10^{-12} up to 10^{-6} mbar and pressure measurements down to 3×10^{-14} mbar are possible.

4.6 SPINNING ROTOR GAUGE (SRG)

It has been realised for many years that the angular deceleration of a spinning ball due to molecular impacts could be used to measure gas pressures in the molecular flow region. However although the first spinning rotor gauge using magnetic suspension was made in 1946 (Beams *et al* 1946) it is unlikely that the SRG would have reached the present sophisticated state of development without the intense efforts of Dr J K Fremerey and his colleagues at KFA, Julich, FRG since about 1971. In a seminal paper Fremerey (1982) described a practical and compact instrument as a result of which a commercial gauge was developed for use in research and industrial laboratories.

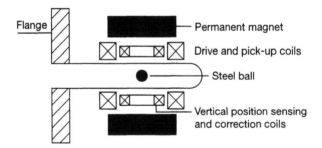

Figure 4.9 Schematic diagram of the spinning rotor gauge.

A schematic diagram of an SRG is shown in figure 4.9. The rotor is a steel ball, typically 4.5 mm in diameter, magnetically levitated inside a stainless steel tube, using a permanent magnet and additional current-generated magnetic fields. The tube is closed at one end and the open end is connected to a flange on the vacuum system under test. The diagram shows the drive and pick-up coils and the vertical sensing and correcting coils. In use the driving magnetic field is rotated and the ball accelerated to about 400 Hz. The driving field is then switched off and the reduction in the rotational speed of the ball, caused by the decelerating effects of gas-induced friction, is measured over a period of time by pick-up coils which sense the rotating magnetic moment. A microprocessor based data processing unit is used to obtain a reading of the pressure which can be up-dated every few seconds using an average measuring period of 10 seconds.

It is shown in appendix A that in the molecular flow region the pressure p is given by the expression

$$p = \frac{\rho d}{10\sigma} \sqrt{\frac{2\pi RT}{M}} \left(\frac{\dot{\omega}}{\omega}\right) - P$$

where ρ and d are the density and diameter respectively of the ball, σ is the tangential momentum transfer coefficient and $(\dot{\omega}/\omega)$ is the measured fractional angular deceleration rate. P is a small offset term representing residual drag effects and is effectively the gauge reading when the pressure is zero. The other symbols have their usual significance. Thus assuming that the gas composition is known then it is only necessary to know σ and P.

The value of the coefficient σ may be slightly greater than unity, by a very few per cent, due mainly to the roughness of the surface of the ball on a microscopic scale (see appendix E). For all but the most accurate work σ may be assumed to be unity (Comsa *et al* 1980). To determine σ values precisely an instrument is calibrated in a known pressure generated in the facilities available at national standards laboratories. The residual reading P is more variable and typically in the range 10^{-7} to 10^{-6} mbar. It is due to decelerating effects of basically electromagnetic origin, mainly eddy current losses which cause braking directly. Indirectly, heating of the ball, with a consequent slight change of moment of inertia due to thermal expansion, can be a problem if the ball has been levitated for times $\sim$ hours.

The normal linear operating range of the SRG is from about 10^{-2} to 10^{-7} mbar with a reproducibility in the readings of better than 1%. The main advantages of the SRG are its accuracy and reproducibility. It can be used with chemically active gases and it produces no pumping effects. Furthermore it is bakeable to 400 °C and because it is a passive device it produces negligible thermal disturbance or excitation of the residual gas molecules. Repeated baking however may cause the magnetic moment to diminish with consequent loss of coupling to the speed sensing coil. If signal loss can be attributed to this and no other reason it may be possible to restore it by passing a weak magnet externally over the tube to remagnetise the ball.

It can be said with complete confidence that the SRG has been the most significant development in the last two decades in the field of low pressure measurement and it is firmly established as a transfer standard (McCulloh *et al* 1985). Improvements in its field of application continue to be made. The SRG has already been accepted in many of the industrial and research laboratories where very careful control of the vacuum environment is necessary, for example in the semiconductor industry.

Lindenau and Fremerey (1991) discuss the use of the gauge at higher than molecular pressures in the transitional and viscous ranges where response becomes less sensitive and non-linear but nevertheless measurably dependent on pressure and gas type.

4.7 CALIBRATION OF VACUUM GAUGES

Calibration of vacuum gauges is an important topic which was carefully described in a paper by Steckelmacher (1987). Its importance has if anything increased in recent years because of the precision required in specifying pressures for processes in semiconductor fabrication. Hinkle and Uttaro (1996) have discussed the methodology and performance of standards for both vacuum and gas flow measurements suitable for the production environment.

There are in principle three alternative procedures by which calibration of a gauge may be achieved.

4.7.1   Comparison of the gauge with an absolute standard which can be calibrated from its own physical properties

It has already been pointed out in section 4.2 that an absolute liquid manometer can be used as a standard from atmospheric pressure to about 0.1 mbar and as low as 10^{-5} mbar if optical interference techniques are used. In this category the McLeod gauge was used for many years, but it was very difficult to achieve accurate and reliable results.

4.7.2   Attachment of the gauge to a calibration vacuum system in which a known pressure can be generated

The known pressure is generated by either a static or dynamic method, depending on the pressure range. A schematic diagram of the *static series expansion technique* is given in figure 4.10 and shows a three-stage expansion. Essentially, Boyle's law is used to determine a final pressure which is the outcome of isothermally expanding a small initial volume of gas at a known pressure ~ 1000 mbar by a known volume ratio.

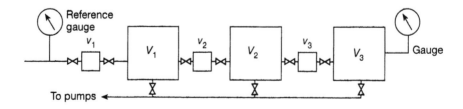

Figure 4.10 Schematic diagram of the series expansion technique.

The large volumes V_1, V_2 and V_3 are connected by the small volumes v_1, v_2 and v_3. The diagram also shows the gauge under test, the reference pressure gauge and the connections to the pumping system. An inert gas such as argon is admitted into v_1 at a known pressure p_1 and the gas is expanded in stages to V_3, via the intermediate volumes. The final pressure, p_3, is given by the equation

$$p_3 = p_1 \times \frac{v_1}{v_1 + V_1 + v_2} \times \frac{v_2}{v_2 + V_3 + v_3} \times \frac{v_3}{v_3 + V_3}$$

All the volumes can be measured independently and therefore p_3 can be determined from the above equation. Depending on the gauges to be calibrated, pressures are generated in a range from 10 down to 10^{-7} mbar, with uncertainties of order 0.1%. For example a pressure of 10^{-6} mbar will have uncertainty 0.2%, higher pressures less. Expansions using inert gases such as argon are less problematic than those using chemically active gases such as oxygen, for which outgassing and adsorption and desorption effects at the vessel walls are significant at the lower pressures. Increasingly the calibration of SRGs in gases of industrial importance is necessary and so this remains an active field. Jitschin *et al* (1990) discuss the method and experimental concerns in detail.

At lower pressures the technique used to generate the known pressure is *dynamic expansion* (also called 'orifice flow').

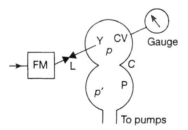

Figure 4.11 Schematic diagram of the orifice flow technique.

A schematic diagram of the orifice flow technique is given in figure 4.11 which shows an upper calibration volume (CV) separated from a pumping chamber (P) by a conductance (C). The gauge to be tested, a flow meter (FM) which measures the throughput Q and a variable leak valve (L) are located in the upper volume, in which the pressure to be determined is p. In the chamber, which is connected to a pump of known large speed S, the pressure is p'. The throughput through the aperture is

$$Q = C(p - p')$$

But $Q = S \times p'$. Hence the pressure p is determined by the equation

$$p = Q(1/C + 1/S)$$

The value of C can either be calculated or determined experimentally. This method can be used down to 10^{-10} mbar.

4.7.3 Comparison of the gauge with a transferable reference gauge which has already been calibrated

Transferable reference gauges are gauges which are suitable for calibration against the primary standards as indicated in the previous section and these are used to check other gauges on various systems and in various locations. The most important transfer gauges are the capacitance manometer, the spinning rotor gauge and certain designs of thermionic gauges (Poulter 1981). However it is necessary that the calibration of these gauges is performed in an appropriate vacuum system as, for example, described by Nash and Thompson (1983).

4.8 PARTIAL PRESSURE GAUGES

It has already been noted that the sensitivity of some gauges varies for different gas species so that if it is important to have an accurate knowledge of the total gas pressure it is necessary to know the composition of the residual gases and the appropriate gauge sensitivities. Furthermore for many applications of vacuum technology it is more important to know the identity of the residual gases rather than the actual value of the total pressure. For example in surface studies it may be necessary to know that the partial pressure of particular chemically active gases in the residual system gas are at or below a certain level, or that others are not present at all. Thus there is a need to incorporate a partial pressure analyser (PPA) into a system. The PPA instrument is more frequently referred to as a residual gas analyser (RGA).

An RGA is essentially a mass spectrometer designed specifically for investigating residual gases in a vacuum system so that it normally has a higher sensitivity but lower resolution than a conventional analytical mass spectrometer, and a more limited mass range - typically from 1 to 100 or 1 to 200 amu. (1 amu = an atomic mass unit is the mass of one atom of hydrogen; the mass of other molecules expressed in this unit is given by the appropriate number in tables 1.1 and 1.2 of chapter 1.) Other matters of concern for an RGA are ease of operation, the need for the signal to be proportional to the partial pressure of the species identified, and that it should be bakeable to at least 250 °C. It is also important that its operation disturbs the system conditions as little as possible.

In an RGA gaseous ions reflecting the composition of the residual gas are generated in an ion source box by electron impact ionisation. They are extracted from it and analysed, then collected and amplified. A parameter such as an applied potential is varied in order to scan the required mass range so that the spectrum of the analysed gas can be displayed on a visual display unit (VDU) or a chart recorder. A typical mass spectrum of an unbaked system is given in figure 4.12. It can be seen that the largest peak is at mass 18 which is that of water vapour.

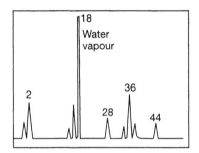

Figure 4.12 Typical mass spectrum of an unbaked vacuum system.

The range of different analyser types which emerged during the early period of their development is described in a review paper by Weston (1980). Devices in which mass separation was based on either magnetic sector or electrostatic quadrupole filtering had come to be the most widely used. Even then the quadrupole instrument was acquiring the position of almost complete dominance which, for reasons to be discussed, it occupies today. However the magnetic method is still used in leak detectors where it has certain advantages and it serves as a useful vehicle for introducing the subject. Outside the RGA application it remains the basis of analytical mass spectrometers used for quantitative chemical analysis.

4.9 THE MAGNETIC SECTOR ANALYSER

There is a number of configurations of magnetic sector spectrometers. The 180° deflection instrument is shown schematically in figure 4.13.

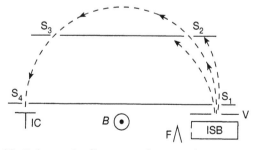

Figure 4.13 Schematic diagram of magnetic sector spectrometer.

An electron beam from a thermionic filament, F, is accelerated into the ion source box (ISB) where some of the residual gas molecules are ionised by these energetic electrons to produce positive ions. These ions are then extracted from the ion source by the negative and variable potential, V, into a uniform magnetic field B, whose direction is normal to the plane of the paper. If the ions are assumed to be singly charged then they will have an energy eV and will travel in a circular path of radius R in the magnetic field through the slits S_1, S_2, S_3 and S_4. If m is the mass and v is the velocity of the ion then the ion energy is

$$eV = \frac{mv^2}{2}$$

and the force Bev on the ion provides the centripetal acceleration so that

$$\frac{mv^2}{R} = Bev$$

whence

$$\frac{m}{e} = \frac{R^2 B^2}{2V}$$

Since R, B and e are constant only those ions with a particular value of m and accelerated through V will pass through all four slits to reach the ion collector (IC). Thus the mass spectrum can be scanned by varying the ion accelerating voltage V, and

$$m \propto \frac{1}{V}$$

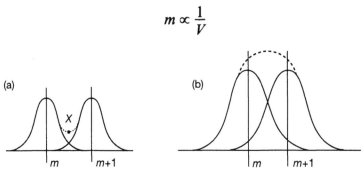

Figure 4.14 Peaks for masses m and $(m+1)$: (a) resolved; (b) not resolved.

The resolving power, $m/\Delta m$, measures an instrument's ability to distinguish between two adjacent mass peaks in the spectrum, and its basis is illustrated in figure 4.14 for mass peaks at m and $(m + 1)$ with equal height. Individual peaks have a width Δm, called the resolution, which is determined by the overall design and operation of the instrument. In figure 4.14(a) the two peaks can be clearly resolved but this is not the case in figure 4.14(b) for which the Δm is greater. In this instrument, as one can imagine by referring to figure 4.13, the resolving power increases with increase of radius R and decrease of slit width. However the sensitivity (defined as ion current per mbar of gas in the source) behaves oppositely - it will increase as R is reduced and slit widths are increased - and so a compromise has to be chosen between resolving power and sensitivity. Of course, the resolving power will also increase with increasing values of m. When point X of figure 4.14(a) is 10% of the main peak height then this is said to be the '10% valley' value. Thus sensitivity may be increased by instrumental adjustments up to a value for which this resolving condition is achieved. In practice this is realised in a 180° magnetic deflection spectrometer with an ion deflection radius of about 1 cm and fixed magnetic field of about 0.2 T making it suitable for use as an RGA with mass range up to about 100 amu.

4.10 THE QUADRUPOLE RESIDUAL GAS ANALYSER

The quadrupole mass filter consists essentially of four cylindrical conducting rods as shown in figures 4.15(a) and (b) to which are applied a combination of DC and RF voltages U and $V\cos(\omega t)$ respectively to produce a rather complex time-dependent electric field in the region between them. The frequency is typically a few megahertz. The ions produced by the ion source pass through the focussing electrodes and enter the quadrupole field as shown in figure 4.15(c). Those ions which pass through the filter produce a current which is collected directly at an electrode or amplified prior to collection. Parameters can be varied so that ions of a particular mass-to-charge ratio pass through the system to be detected, whilst all others are filtered out, being collected at the rods.

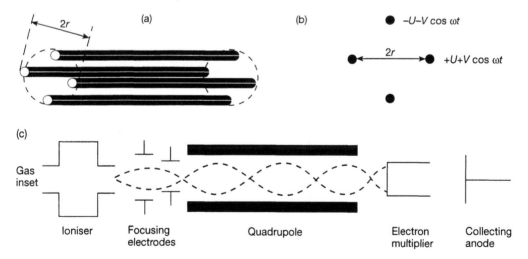

Figure 4.15 (a) The quadrupole rods. (b) Application of DC and RF voltages. (c) Overall arrangement of the quadrupole.

The potential ϕ at any point in the quadrupole field can be written as

$$\phi = (U + V\cos\omega t)(x^2 - y^2)/2r^2$$

where ω is the angular frequency and $2r$ is the rod spacing. The resulting differential equations of motion for an ion accelerated into the field in the x, y and z directions can then be set up. The equation in the axial z direction shows that the acceleration is zero and therefore the ion continues to travel with its original velocity component in this direction. The equations in the x and y directions - known as the Mathieu equations - have been solved and give stability diagrams in terms of dimensionless parameters q and α defined below. A stable ion trajectory is one for which the ion passes through the system rather than being lost by moving radially outwards to be collected by the quadrupole rods. A plot of q against α as shown in figure 4.16 where, for an ion of mass m

$$q = 2eV/m\omega^2r^2 \quad \text{and} \quad a = 4eU/m\omega^2r^2$$

The stability diagram is used to optimise the operating parameters a, q, U, V, ω and r so that ions with a particular charge-to-mass ratio within a narrow range describe stable trajectories and pass through the system, whereas others follow unstable trajectories and are collected by the rods. The highest resolution is achieved near the apex of the diagram and may be kept constant over the complete mass range. If ω and the ratio U/V is kept constant, the mass-to-charge ratio is proportional to the amplitude V. The mass scan, which is linear, is often done by varying U and V but keeping U/V and ω constant. Alternatively the mass scan may be done by keeping U and V constant and varying ω. The injection of the ions into the quadrupole is not critical and ions which are at even 20° to the axis can be accepted, but care must be taken to avoid fringing fields near the end of the quadrupole rods.

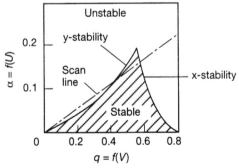

Figure 4.16 Stability diagram of quadrupole spectrometer.

In recent years there has been a considerable development of small quadrupole RGAs for use in scientific and industrial vacuum environments. They are now available covering the mass ranges 1 - 100 or 1 - 200 amu and in some cases 1 - 300 amu. They are compact devices about 150 mm long overall and 35 mm in diameter mounted on 70 mm (2¾ inch) diameter standard flanges with head electronics housed in a small box on the atmospheric side of the flange. Typical minimum detectable partial pressures are ~10^{-11} mbar with a simple Faraday cup type of ion current collector, and ~10^{-13} or less when an electron multiplier is used. These amplify the current by exploiting the secondary electron emission phenomenon in either a channel electron multiplier or a dynode chain, as described by O'Hanlon (1989).

Quadrupole instruments cannot be operated at pressures much higher than 10^{-5} mbar because of their thermionic filaments. Also at these higher pressures electron multipliers if fitted should not be inadvertently (and unnecessarily) left on because they will be overloaded and likely damaged. It is however a frequent industrial need to do gas analysis at much higher pressures than this. Sometimes this problem can be solved by housing a quadrupole in a auxiliary pumped high vacuum chamber and taking a continuous sample of gas into it from the system at higher pressure via a suitably designed connection of known conductance for the various gases sampled. Gas analysis is thereby accomplished while keeping the instrument under its proper operating conditions.

4.11 DISPLAY OF SPECTRA

An important feature of the quadrupole is that, especially when equipped with an electron multiplier, it can be used at very fast scan rates. This makes it very compatible with modern electronics so that the instrument can be put under the control of a desk-top computer. This increases the versatility of the instrument considerably in terms of the variety of displays of output information that are possible, as is described below. These and many other aspects of the quadrupole were well documented in papers by Batey (1987) and James (1987).

Most manufacturers supply 'Windows' or other software for controlling the instrument. Basically this drives it, acquires data from it and stores and processes that information. The basic information acquired is the spectrum of peaks at different masses as the instrument scans the mass range. When presented in *analogue mode* the peak shapes are evident, and adjustments from the keyboard can be made, if desired, to resolution, sensitivity and the range scanned. In displaying peak height versus mass number the scale of the former may be linear, i.e. directly proportional to their true magnitude as in traditional instruments, or logarithmic with a range of up to six decades. The logarithmic scale can be very useful in indicating the presence of gases in very small amounts. Thus a peak for gas X which is one-third of the size of a peak for gas Y indicates that the amount of X is in fact only 0.1% of Y. In revealing many peaks by boosting weak signals the logarithmic display can sometimes present information overload at first sight, and switching to a linear or reduced logarithmic range reveals a simpler display which emphasises only the main features present. The *histogram* (or *bar-graph*) *mode* of the same information of peak height versus mass number, again with a choice of vertical scale sensitivity, looks 'cleaner' than the corresponding analogue mode and often allows easier recognition of groups of features.

The *library mode* may contain the stored spectra of a wide range of gases and vapours which can be compared, using a split-screen facility, with an acquired spectrum under investigation. The *search mode* may be used to initiate analysis of a spectrum for which there is some prior knowledge of likely composition.

Particularly valuable is the *pressure versus time mode* in which the magnitude of up to ten mass peaks can be monitored and displayed, each having a different colour, with a time scale whose span across the screen, minutes or hours, is set to suit the purpose in hand. The correlation or lack of it between the signals is very informative in helping to identify the source of some operational problems. This mode has many applications, for example in following the changes to gas composition that might be caused by local heating when an evaporation source is switched on, or changes occurring during an overnight baking period.

In the *leak detector mode* a partial pressure of a test gas, frequently helium, may be monitored in time as the gas is presented on the atmospheric side of the vacuum wall. The signal increase indicating a leak can be announced sonically as well as on the screen.

Other modes available are the *table mode* giving tabulated screen displays of selected data and the *annunciator mode* in which the screen is divided into large boxed areas, easily seen from a distance of a few metres, with each box specific to a particular mass and colour coded red or green, say, to indicate the status, above or below a critical process level, of components in the gas mixture.

The controlling software can usually take data from more than one quadrupole head enabling conditions to be monitored and compared in different locations, for example a main vacuum chamber and an auxiliary connected region such as a load-lock.

4.12 INTERPRETATION OF SPECTRA

When the energetic electrons in the ion source of the RGA collide with gaseous molecules most of the ions produced are simply those of the parent molecule modified by the loss of just one of its orbital electrons. But in addition the energetic electrons will, to some extent, break the molecule apart into fragments and form ions of them. The resulting mass spectrum is usually referred to as the 'cracking pattern' of the molecule; each fragment has a particular value of the ratio m/e and a corresponding peak in the spectrum. Such fragments are generated in the ion source box and are not present in the system's residual gas. Water vapour for example gives peaks at masses 1, 16 and 17 which in magnitude are typically 3, 3 and 27% respectively of that of the main peak at mass 18. They are associated with H^+, O^+ and OH^+ ionised fragments of H_2O.

If there are two gases present then each will have its own characteristic cracking pattern, and we can only determine the proportions of each gas from the relevant peak heights if we know their cracking patterns, either from data compilations or ideally, since instruments differ in their detailed sensitivities, perhaps from previous calibrations of the instrument in use. In practice the analysis also involves other factors which have to be taken into account in the interpretation of spectra and a few are discussed briefly below. For a thorough and practical introduction to the problems which arise in analysing spectra the reader should consult O'Hanlon (1989).

(1) *Presence of multiply-charged ions.* Because there is the possibility of some nitrogen molecules, for example, losing *two* electrons as a result of being struck by an energetic incident electron, molecular nitrogen has two peaks at different values of m/e namely:

$$N_2^+ \text{ at } m/e = 28 \quad \text{and} \quad N_2^{++} \text{ at } m/e = 14 \ (9\% \text{ of the 28 peak})$$

(2) *Different gases having the same value of m/e.* For example, molecular nitrogen (N_2) and carbon monoxide (CO) which have the same molecular mass (2 ×14 and 12 + 16 respectively) will both give signals at $m/e = 28$. Their contributions can only be distinguished by looking at the cracking patterns. Thus nitrogen will have an associated peak at $m/e = 14$ corresponding to the presence of N_2^{++} while for CO there will be peaks at 12 and 16 corresponding to C_{12}^+ and O_{16}^+ (approximately 3.5 and 1.5% of the 28 peak).

(3) *Different masses having the same value of m/e.* For example singly ionised neon Ne_{20}^+ with $m/e = 20$ and doubly ionised argon Ar_{40}^{++} at $m/e = 20$.

(4) *The existence of isotopes.* For example the nitrogen atom exists as two isotopes of mass 14 amu (99.63%) and 15 amu (0.63%). Therefore there will be nitrogen molecules with mass 28, 29 or 30, though mass 28 will be dominant.

Very useful tabulations of the cracking patterns of common gases and vapours, of the vapours of pump fluids and of the isotopic abundances are given in O'Hanlon's book already mentioned.

A simple but important example of a residual gas spectrum was shown in figure 4.12 which is typical of a vacuum system at a pressure of about 10^{-6} mbar that has not been baked. The spectrum is dominated by water vapour. Residual gas spectra of representative vacuum systems are to be found in many texts, notably in the AVS monograph of Drinkwine and Lichtman (1995), which covers the whole subject of partial pressure analysis in a thorough and accessible way.

It is arguable that an RGA is an essential part of a vacuum system even if one is not concerned to have a detailed analysis of the residual gases, because the residual gas spectrum may give important and otherwise unknown information about the state of the system. The information may not always be welcome, but it may not be irrelevant! Of course an in-built RGA has the merit that it can also be used as a leak detector.

Summary
Various types of total and partial pressure gauge have been described, the selection being based mainly on the criterion of their widespread use in appropriate roles. Table 4.1 indicates for quick reference the ranges of pressure in which they can be used.

Broadly speaking thermal conductivity gauges monitor the low and medium vacuum regions without great precision but at relatively modest cost. Capacitance manometers offer much greater accuracy over this range. The spinning rotor gauge offers high accuracy in the medium and high vacuum range. The high and ultra high vacuum regions are served by hot and cold cathode ionisation gauges which, with the exception of the Penning gauge, may be expected to be accurate to $\pm$ 10 - 20% if carefully used.

The review article by Tilford (1992) gives a broad survey of principles and practice in pressure measurement with substantial discussion of the characteristics and limitations of individual types of gauge, and valuable advice on operational matters based on extensive investigations carried out at the US National Institute of Standards and Technology.

Table 4.1 Typical ranges of a selection of gauges.

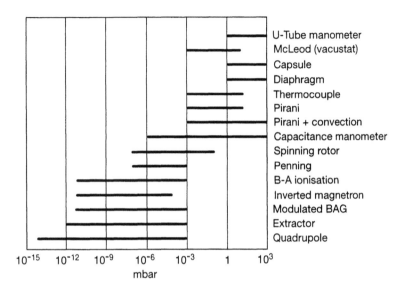

References

Akimichi H, Arai T, Takeuchi K, Tuzi Y and Arakawa I 1997 *J. Vac. Sci. Technol.* A **15** 753

Akimichi H, Takahashi N, Tanaka T, Takeuchi K and Tuzi Y 1996 *Vacuum* **47** 561

Akimichi H, Tanaka T, Takeuchi K and Tuzi Y 1995 *Vacuum* **46** 749

Arnold P C and Borichevsky S C 1994 *J. Vac. Sci. Technol.* A**12** 568

Batey J H 1987 *Vacuum* **37** 659

Beams J W, Young J L and Moore J W 1946 *J. Appl. Phys.* **17** 886

Beeck U and Reich G 1972 *J. Vac. Sci. Technol.* **9** 126

Benvenuti C and Hauer M 1980 *Proc. 8th Inst. Vac. Cong.* (*Cannes*) vol 2, p 199 (Paris: Soc. Francaise du Vide)

Berman A 1985 *Total Pressure Measurement in Vacuum Technology* (New York: Academic)

Bigelow W C, 1994 *Vacuum Methods in Electron Microscopy* (London: Portland)

Comsa G, Fremery J K, Lindenau B, Messer G and Rohl P 1980 *J. Vac. Sci. Technol.* **17** 642

Drinkwine M J and Lichtman D 1995 *Partial Pressure Analysers and Analysis* AVS Monogr. Ser. (Woodbury, NY: AIP Press)

Durakiewicz T and Halas S 1995 *Vacuum* **46** 101

Fitch R K 1987 *Vacuum* **37** 637

Fremery J K 1982 *Vacuum* **32** 685

Helmer J C and Hayward W H 1966 *Rev. Sci. Instrum.* **37** 1652

Hinkle L D and Uttaro F L 1996 *Vacuum* **47** 523

Hobson J P and Redhead P A 1958 *Can. J. Phys.* **36** 271

James A P 1987 *Vacuum* **37** 677

Jitschin W, Migwi J K and Grosse G 1990 *Vacuum* **40** 293

Kendall B R F and Drubetsky E 1997 *J. Vac. Sci. Technol.* A15 740

Leck J H 1970 *Vacuum* **20** 369

Leck J H 1989 *Total and Partial Pressure Measurement in Vacuum Systems* (Glasgow and London: Blackie)

Lindbland Th, Bagge L, Bjon J and Leven S 1987 *Vacuum* **37** 293

Lindenau B 1988 *Vacuum* **38** 893

Lindenau B E and Fremerey J K 1991 *J. Vac. Sci. Technol.* A9 2737

McCulloh K E, Wood S D and Tilford C R 1985 *J. Vac. Sci. Technol.* A3 1738

Nash P J 1987 *Vacuum* **37** 643

Nash P J and Thompson T J 1983 *J. Vac. Sci. Technol.* A1 172

OHanlon J F 1989 *A User's Guide to Vacuum Technology* (New York: Wiley)

Peacock R N Peacock N T and Hauschulz D S 1991 *J. Vac. Sci. Technol.* A9 1987

Poulter K 1981 *Vide-Couches Minces* **36** 521

Redhead P A 1993 *Vacuum* **44** 559-564

Redhead P A 1987 *J. Vac. Sci. and Technol.* A5 3215

Redhead P A, Hobson J P and Kornelsen E V 1968 *The Physical Basis of Ultra-high Vacuum*, reprinted (1993) in AVS Classics Ser.(Woodbury NY: AIP Press)

Steckelmacher W 1987 *Vacuum* **37** 651

Sullivan J J 1985 *J. Vac. Sci. Technol.* A3 1721

Tilford C R 1992 *Physical Methods of Chemistry* 2nd edition, eds B W Rossiter and R C Baetzold (New York: Wiley) pp 101-73

Tilford C R, Filippelli A R and Abbott P J 1995 *J. Vac. Sci. Technol.* A13 485

Watanabe F 1993 *J. Vac. Sci. Technol.* A11 1620

Watanabe F 1996 *Vacuum* **47** 567

Weston G 1980 *Vacuum* **30** 49

Weston G 1985 *Ultrahigh Vacuum Practice* (London: Butterworth)

Wutz M, Adam H and Walcher W 1989 *Theory and Practice of Vacuum Technology* trans. W Steckelmacher (Braunschweig: Vieweg)

5

Vacuum Materials and Components

B S Halliday

The overall design of all vacuum equipment is essentially determined by the nature and size of the processes involved and the ultimate pressure required. However the detailed design will be influenced by the suitability and range of the materials and components available and thus the correct choice of these is essential if performance and cost are to be optimised. Of course it is also of prime importance that any material or component used must not prevent the attainment of the degrees of vacuum dictated by the process, nor disturb the vacuum process or create a hazard to the health and safety of the operators. It is therefore clear that the construction, materials, components and fluids play an integral part in the performance of any vacuum system. Sections 5.1 - 4 discuss materials, 5.5 fluids, 5.6 - 7 components and 5.8 their construction.

5.1 PROPERTIES REQUIRED
The material must be able to be machined and fabricated to produce the component, adequately strong at the maximum temperature of operation and must not become brittle at the lowest operating temperature. Furthermore the material should retain its elastic, plastic and fluid properties at the extremes of operating temperature and pressure and sometimes in the presence of radiation.

5.1.1 Thermal
The vapour pressure must remain adequately low at the highest operating temperature and the thermal expansion of adjacent materials must match and not cause any distortion or change of tolerance (e.g. flange bolts in UHV systems which undergo bakeout).

5.1.2 Gas loading
The material must be impermeable to gases, must not be porous or have any cracks or crevices as these can create virtual leaks, trap dirt and retain cleaning fluids. After the initial treatment and before use (cleaning, heat treatment etc) the surface and bulk desorption rates must be adequate at the extremes of temperature (and radiation).

5.1.3 Reactions
The materials selected must not react with any other materials in the system and must not react with the vacuum process.

5.1.4 Radiation
The materials must not evolve excessive gas when subject to irradiation from neutrons, x-rays or high energy particle beams, or degrade so as to be inadequate (e.g. elastomers hardening and view port glasses darkening).

5.2 COMMONLY USED MATERIALS

5.2.1 Metals
Austenitic stainless steels.
Advantages: types generally preferred are EN58A, EN58B (US321), EN58E (304) and EN58F (347) with EN58B and EN58F chosen most frequently for satisfactory welding by an argon-arc. If low magnetic permeability required, EN58B is used. All have a low outgassing rate as shown in figure 5.1.
Disadvantages: EN58F will not accept a high polish and welding can cause considerable distortion in all grades.

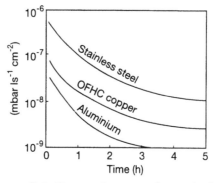

Figure 5.1 Outgassing rates for typical metals.

Aluminium and aluminium alloys.
Advantages: aluminium and magnesium alloys are most used, but a high zinc content should not be used. Good corrosion resistance, easily machined and jointed.
Disadvantages: strength at high temperatures is poor. Alloys with copper content present welding problems. High distortion when welding which may lead to further machining.

Nickel alloys (Inconel, Kovar).
Advantages: high strength at high temperatures, excellent corrosion resistance.
Disadvantages: not readily available, high cost and present machining problems.

Copper.
Advantages: oxygen free, high conductivity grade (OFHC) excellent for vacuum systems, easily machined, good corrosion resistance.
Disadvantages: brazing in a hydrogen atmosphere sometimes difficult.
Brass.
Advantages: suitable for some exceptional applications, care must be taken in grade selection, good corrosion resistance. The use of castings should be given careful consideration to avoid virtual leaks.
Disadvantages: brass contains zinc (15 - 20%). Zinc evaporates out at temperatures over 100 °C.
Mild steel.
Advantages: may be used generally down to 10^{-3} mbar, can be used at lower pressures if plated after welding.
Disadvantages: liable to rust.
Titanium. A good clean metal in vacuum, light in weight and ductile. Used mostly for its gettering properties e.g ion pump cathodes, and getter pump filaments.

5.2.2 Plastics

Plastics generally desorb large quantities of gas and have a high permeability rate compared with metals, so that they must be carefully considered and generally kept to a minimum. Outgassing rates for some plastics are given in figure 5.2.
PTFE. Low outgassing rate, good electrical insulator, can be used at a higher temperature than most plastics, self-lubricating.
Glass-filled PTFE - a form of PTFE which has less tendency to cold flow.
Polycarbonate. Moderate outgassing rate and water absorption, good electrical insulator.
Nylon and acrylic. High outgassing and water absorption rates, self-lubricating.
PVC. High outgassing and water absorption rates, flexible PVC tubing useful for backing lines and temporary connections (e.g. leak detectors).
Polyethylene.   Only suitable if well outgassed.
PEEK is a high temperature polymer, useful as an insulator where low temperature bakeout is used.
Electrical wires. Electrical wires coated with Kevlar have good electrical insulation. Ceramic beads are also useful and prevent trapped volumes.

5.2.3 Synthetic resins
Not recommended.

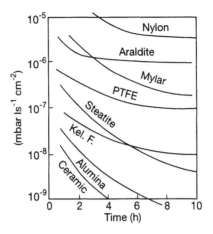

Figure 5.2 Outgassing rates for typical plastics and glasses.

5.3 SEALS

5.3.1 Elastomer seals

Nitrile rubber.   The most commonly used sealing ring is nitrile rubber which can be easily jointed (*in situ* for large installations).

Viton.   Is most suitable for seals at lower pressures, has low outgassing rates and is heat resistant. It has a tendency to remain deformed for some time after compression. New viton should be vacuum baked at 100 °C for 1 hour to remove moulding release agents.

PTFE.   Has a very poor compression set and is not generally recommended.

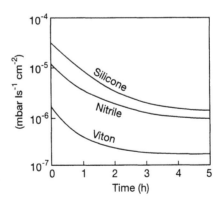

Figure 5.3 Outgassing rates for plastics often used for demountable seals.

5.3.2 Metal seals

Copper ring.   Used with knife edges machined into opposing flanges (e.g. Conflat). Can be baked up to 450 °C. Care must be taken that the knife edges do not become damaged in store or during installation.

Aluminium diamond edged disc seal. Fits between flat machined flanges. Self-aligning, bakeout to 200 °C. Flange finish required - 0.8 μm.

'Wills' rings. Can be used at high temperature and pressure. Needs higher flange bolt loading than knife edge and requires a higher flange finish than the diamond edge seal.

Indium wire. Very soft, continues to flow after initial tightening of flanges. Can be easily re-extruded.

Aluminium wire. Easy to manufacture from annealed wire by electric butt-welding. Fits between flat machined flanges. Requires aluminium foil centring tabs. Flange finish required - 0.8 μm.

Gold wire. Inert to all gases, bakeout to 450 °C. Needs high flange bolt loading and well finished flange faces (0.4 μm). Expensive initially but good cost recovery for scrap.

5.3.3 Proprietary brand seals

Essential to follow manufacturers' tightening instructions. Bake-out up to 450 °C. (Vaculok, Swagelock, Betabite etc.)

5.4 CERAMICS AND GLASSES

5.4.1 Ceramics

Fully vitrified electrical porcelain and vitrified alumina are most suitable for vacuum components and insulators and can be used up to a temperature of 1500 °C. Care must be taken in handling these materials as they are brittle and the surfaces mark easily, so they should be handled wearing lint free gloves. Selected ceramics can be brazed to suitable metals by the use of a molybdenum - manganese fired coating onto the ceramic. Machinable ceramics are available and are satisfactory for specific components. They may be machined, drilled etc following manufacturers' instructions.

5.4.2 Glass

Can be used for constructing vacuum systems or parts of systems. The most generally used is borosilicate glass (e.g. Pyrex) which can be obtained in matching components from stock. Glass is also used for viewing windows and selected optical glass can be used for the transmission of ultra-violet light. The permeability of glass varies with temperature and glass has a high corrosion resistance.

5.5 PUMP FLUIDS

5.5.1 Rotary pump fluids

These fluids are usually selected from high quality mineral oil with a low vapour pressure and having good lubricating properties. When it is necessary to pump aggressive or corrosive gases or vapours, the perfluoro polyether derived fluids are used (e.g. Fomblin) as these fluids are generally inert, and have good chemical stability and oxidation resistance.

5.5.2 Diffusion pump fluids

There are four main groups of fluids used.

Hydrocarbon oils. These are prepared by molecular distillation from mineral stock and are suitable for general applications down to 10^{-7} mbar. They will not withstand repeated admission of air when hot as in a quick cycling plant and will produce carbonaceous compounds with high vapour pressures which will degrade pump performance. Any deposits on surfaces in systems which are under charged particle bombardment will produce conducting layers.

Silicone fluids. These are exceptionally stable fluids at high temperatures and will provide ultimate pressures between 10^{-5} and 10^{-9} mbar. They are useful in industrial plant where quick cycling is required. Breakdown of vapours on electrode surfaces due to bombardment will produce electrically insulating layers which makes them unsuitable for charged particle applications (e.g. electron microscopes). The fluids are poor lubricants.

Polyphenyl ethers (e.g. Santovac5*, Convalex10**). These fluids have exceptionally low vapour pressure, are thermally stable and can be used in clean systems (electron microscopes, surface physics studies). They will not easily back-stream in a well designed pump and baffle, and are chemically stable. Breakdown products on surfaces due to bombardment are electrically conducting and they have good lubricating qualities. They are however an expensive fluid. (Trade names of *Monsanto products, **CVC products.)

Perfluoro polyether (e.g. Fomblin***). This fluid is chemically stable and has high oxidation resistance. It is used where energetic particle bombardment would normally form a polymer (with the exception of some hydrogen ions). It is also inert to many aggressive chemicals including oxygen, halogens and mineral acids. It is used in electron microscopes and in ion implanters. It is an expensive fluid and has generally a lower pumping speed for air than other fluids. (***Trade name of Montedison SpA).

5.5.3 Greases

These are generally mineral oil, silicone or Fomblin based and are used to lubricate moving shafts where they enter the vacuum system. Greases should only be used sparingly and are unnecessary on static flange seals if the ring and flange finishes are unimpaired. Some high viscosity pump fluids can be used as seal lubricants.

5.6 FLANGES

These are generally designed to International (ISO) or National (DIN, ASA, Pneurop) standards which regulate dimensions, number, diameter and pitch circle of holes or number of clamps required and the flange thickness. Non-standard flanges may be required for special cases. Flanges can be divided into three groups.

5.6.1 Small flanges

These are 10 - 50 mm bore with a clamp or screwed connecting element, and are used for connecting small bore pipes, small valves, gauges and flexible elements, and enable these connections to be made quickly and precisely. They chiefly use elastomer seals, but special aluminium or indium alloy seals can be used, the choice being governed by the working pressure and temperature as shown in table 5.1.

Table 5.1 Working temperature of seals.

Seal	Working temperature (°C)
Neoprene	90
Indium alloy	90
Viton	150
Aluminium	200

Typical examples of these small flanges are shown in figures 5.4(a) and (b).

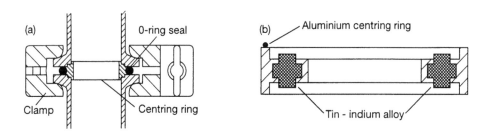

Figure 5.4 Typical small flanges and seals. (a) Klein flange (KF) and (b)Tin/indium alloy seal.

5.6.2 Large flanges

These are 6.3 - 1000 mm bore and can be fastened together by bolts or external clamps and use elastomer or soft wire sealing elements. The number of bolts and flange thickness varies with diameter and some typical examples are shown in table 5.2.

Table 5.2 Large flange parameters.

ISO Flange ref. No	Bore (mm)	No of bolts	Flange thickness (mm)
63	70	4	12
160	153	8	12
250	261	12	12
500	501	16	17
1000	1000	32	24

Flanges may be fixed to tubes or are rotatable to give ease of alignment. Elastomer seals are in grooves or supported between the flanges by a plastic or metal centring ring. These latter have advantages as the flanges are sexless and easier to machine and maintain. Examples are shown in figures 5.5(a) to (g).

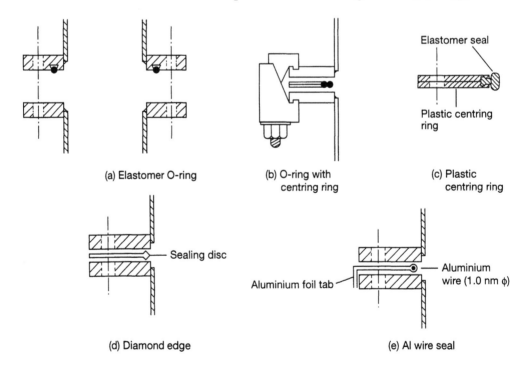

(a) Elastomer O-ring

(b) O-ring with centring ring

(c) Plastic centring ring

Elastomer seal

Plastic centring ring

(d) Diamond edge

Sealing disc

Aluminium foil tab

(e) Al wire seal

Aluminium wire (1.0 nm φ)

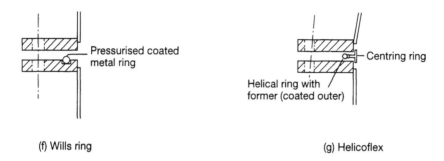

(f) Wills ring   (g) Helicoflex

Figure 5.5 Examples of flange seals.

5.6.3 UHV flanges

Flanges for UHV are nearly always subject to some degree of bake-out. The most commonly used flange is the Conflat* configuration. These are constructed from stainless steel and use an OFHC copper ring. The copper ring is clamped between two knife edged flanges which bite into the copper and force it axially and radially outwards, the flange preventing continued flow. This is shown diagramatically in figure 5.6.

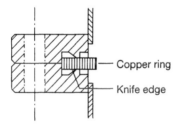

Figure 5.6 A cross-section of a Conflat flange.

The seal can be baked up to 450 °C for UHV purposes. It is used not only for UHV applications but also when deterioration of an elastomer ring would occur due to corrosive or volatile gases, and when nuclear radiation is present. A special viton seal can be used for some room temperature applications and is useful for initial assembly and leak testing as the copper ring can be used once only. These flanges can be fixed or rotatable. The copper sealing rings can be obtained silver plated. This will prevent oxidation of the copper exposed to the atmosphere during bake-out and the subsequent dropping of the oxide particles into the vacuum system when seals are broken during alterations or modifications. Conflat flanges are in standard sizes 64 mm to 200 mm nominal bore.

(* Trade name of Varian, Palo Alto, USA.)

5.7 FITTINGS

Most manufacturers of vacuum components make standard pipe lengths, elbows, T-pieces, cross-pieces and bellows with flanges to order. These are available in the small (KF and screwed) and large flange ranges.

5.7.1 Bellows

These are usually made from stainless steel tubing and are in two forms.

Convoluted bellows. Convoluted bellows, as shown in figure 5.7, are made by hydraulic forming but they are stiff with a high spring rate, and lateral displacement is only possible over long lengths. These bellows are also made as metal hose up to 1000 mm in length and often of tombak metal.

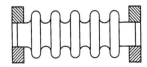

Figure 5.7 Convoluted bellows.

Edge welded bellows. Edge welded membrane bellows are made from stamped and formed stainless steel discs, edge welded internally and externally, as shown in figure 5.8.

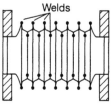

Figure 5.8 Edge welded bellows.

They have generally a lower spring rate and permit some lateral displacement. They can be used as vibration isolators in pipework and are frequently used to seal the actuators of valves and to allow linear motion for manipulators.

5.7.2 Feed-throughs

These are used where it is necessary to transmit movement, fluid, electrical supplies or signals through the walls of a vacuum chamber.

Liquid. These are used for cooling or heating components within the work chamber by circulating fluids.

Electrical supplies or signals. Suitably designed glass, ceramic or porcelain insulators, soldered or brazed to a flange supporting sealed conductors. Signals can be transmitted by the same method using multi-pin and co-axial connections, as shown in figure 5.9.

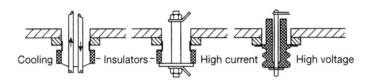

Figure 5.9 Examples of cooling and electrical feed-throughs.

Linear motion. Linear motion shafts are sealed by vacuum greased elastomer rings or discs, or by the flexing of edge welded bellows as shown diagrammatically in figure 5.10.

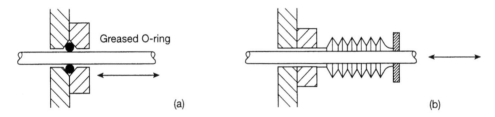

Figure 5.10 Linear motion seals. (a) Greased elastomers. (b) Bellows sealed.

Rotary motion. Rotary motion shafts are sealed by a vacuum greased elastomer O-ring or disc as shown diagrammatically in figure 5.11(a).

Wobble shaft. This is where an offset end to the shaft rotates in a fixed bearing cup contained in the sealed end of a flexible bellows. This is only suitable for slow (hand) turning as shown in figure 5.11(b).

Magnetic coupling. Two permanent magnets are located close to each other but separated by a non-magnetic stainless steel diaphragm. When the external magnet is rotated, the internal magnet follows. This is also only suitable for slow turning and has limited torque transmission as shown in figure 5.11(c).

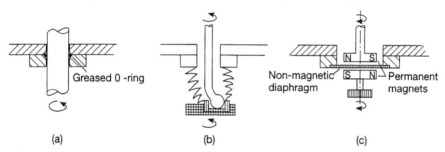

Figure 5.11 Diagrams of rotary feed-throughs. (a) Greased elastomer. (b) Wobble shaft. (c) Magnetic coupling.

A *Ferrofluidic** rotary feed-through, where the seal is made by points sealed with ferric particles held suspended in a suitable low vapour pressure fluid in a strong permanent magnetic field. Suitable for all speeds (0 - 50 000 rpm) and torques. Figure 5.12 shows a diagrammatic section of a ferrofluidic seal. (*Ferrofluidics Corp., Nashua, USA.)

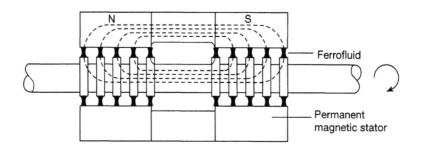

Figure 5.12 A ferrofluidic seal.

5.7.3 Viewing windows
The glass is sealed by a transition metal flange fused to it and brazed to the main flange. The glass can be borosilicate for simple viewing or where optical quality is required a ground sapphire is used. It may be necessary to use a radiation resistant glass in certain cases (e.g. particle accelerators). In coating and sputtering applications internal shutters are often used to prevent the window being coated during deposition. Figure 5.13 shows a cross-section of a window.

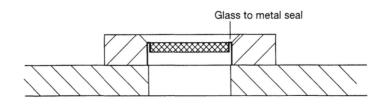

Figure 5.13 A cross-section of a viewing window.

5.7.4 Manipulators or goniometers
These are used for remote positioning of targets and samples inside the vacuum chamber. They employ sliding linear movements coupled with rotary angular movements as shown in figure 5.14. In both cases, cold welding is a problem which can be prevented by the use of dissimilar metals at sliding and mating surfaces - stainless steel and beryllium copper are two suitable metals. The use of powder lubricants such as molybdenum disulphide, tungsten diselenide or boron nitride can be very effective.

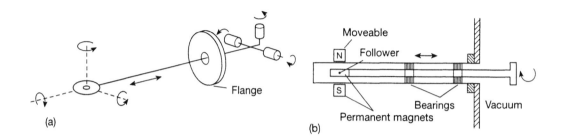

Figure 5.14 (a) A diagram showing motions of a manipulator; (b) magnetic manipulator.

Another form of manipulator consists of a flanged tube with a sealed end, containing a rod with a permanent magnet fixed to one end. An external magnet will couple with the internal magnet to produce lateral and rotational movement inside the vacuum.

A stepping motor for drives within a vacuum system has been developed by AML Microelectronics Ltd of Arundel UK. It removes the need to have a rotary feedthrough through the vacuum wall and allows freedom to place the motor in the most effective position in relation to the driven mechanism. The motors are UHV compatible, have no trapped volumes, are bakable to 200 °C and are usable at 10^{-10} mbar.

5.7.5 Valves
Diaphragm valve. A flexible membrane, figure 5.15, is forced down by a screw to close the aperture. It has a fairly high outgassing rate due to the large area of flexible material.
Uses: Roughing and vacuum lines down to 10^{-3} mbar.

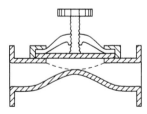

Figure 5.15 A diaphragm valve.

Right angle valve. The seal is achieved by pressure of an O-ring sealed plate actuated by a screw whose shaft is sealed with an O-ring or bellows. Up to 40 mm bore, manual or solenoid operated.
Uses: Roughing lines, venting valves, leak test valves, as shown in figure 5.16.

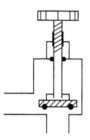

Figure 5.16 A right angle valve.

All-metal right angle valve. The seal is made by a knife edge penetrating a copper surface in the closing plate, figure 5.17. The shaft is bellows sealed, and this valve can be baked. Manual operation only. Uses: UHV systems up to 35 mm bore.

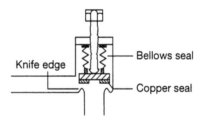

Figure 5.17 An all-metal right angle valve.

Butterfly valve. This has a disc with an O-ring edge seal rotating through 90° in a solid bore. It has good conductance when open, figure 5.18. The O-ring seal must be lubricated.

Uses: Diffusion pump inlet control. It can have manual, pneumatic or motorised control and can be used as a throttle valve.

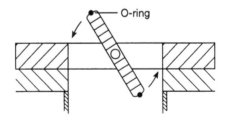

Figure 5.18 A butterfly valve.

Variable leak valves.

Needle. A long tapered needle is moved in an opening increasing or reducing the annular gap. The control is generally manual by an external lever, figure 5.19. The needle is sealed by a bellows.

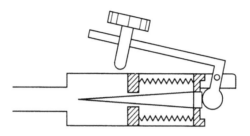

Figure 5.19 A needle valve.

UHV fine leak valve. This has a valve seat of a copper knife edge and a valve plate of sapphire set in a flexible membrane. It can be moved by an external lever giving fine adjustment of the gap, figure 5.20. The assembly can be baked to 450 °C and the range of control is from 500 to 10^{-10} mbar l s^{-1}.

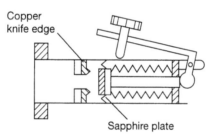

Figure 5.20 A UHV fine leak valve.

Piezo-electric. Variations in applied voltage to a crystal deform the crystal in proportion to the voltage, opening or closing an orifice. This device lends itself to servo control in conjunction with a pressure measuring element.

Gate valve. This valve has a high conductance when open as it leaves no obstruction in the valve body opening, figure 5.21.

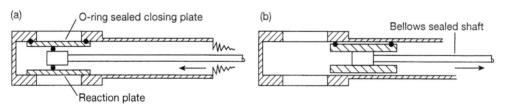

Figure 5.21 A gate valve in the closed (a) and open (b) positions.

The valve plates expand on closing and lock in the closed position by a toggle mechanism. The plates contract inward before withdrawing to prevent damage to the O-ring seal. These valves can be operated manually or frequently by an electro-pneumatic piston. Sizes are between 16 and 1350 mm bore.

Uses: Gate valves are used where clear passage is required as in particle accelerators, electron microscopes and in loading locks to introduce samples into a vacuum chamber.

Fast-acting valves. These are most frequently used to protect systems which have extensions with fragile components (e.g. foil windows) and whose incorrect operation can have long term effects. They are usually spring operated with an electromagnetic release mechanism triggered from a remote pressure sensor, figure 5.22. The closing times vary depending on valve size e.g. a 63 mm bore closes in 10 ms and a 250 mm bore closes in 50 ms.

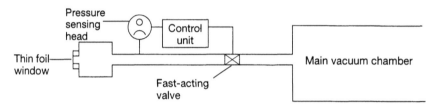

Figure 5.22 Schematic diagram of a fast-acting valve application.

5.8 MACHINING AND CONSTRUCTION OF COMPONENTS

5.8.1 Direction of machining marks
Where the machining of a component can be accomplished in more than one way, e.g. different set-ups on a milling machine, consideration should always be given to the method which, in addition to giving the best possible finish, will also leave any machining marks parallel to and not across the run followed by the gasket. This should be indicated on a drawing by either circular lay or by the direction of lay being shown.

5.8.2 Surface finish
Particular attention should be paid to surface finish (1) to avoid leak paths where the machining marks cross O-ring seals etc, and (2) to facilitate cleaning by ensuring that grease or dirt does not remain on any rough surface resulting from the coarser types of machining.

5.8.3 Marking out
When marking out, ensure that there are no scribed lines on the vacuum sealing surfaces, thus avoiding the possibility of introducing a leak.

5.8.4 Protection
Immediately after sealing surfaces have been finished they should be wrapped in aluminium foil and a protective blank (such as stiff cardboard) should be firmly

affixed to protect the surface against accidental damage. The item should then be sealed in a polyethylene bag. Mild steel surfaces of components being sent to store must be adequately protected to avoid rusting. Sealing surfaces should be inspected and cleaned before use. (See section 6.4.)

5.8.5 Trapped volumes

To avoid trapped volumes, i.e. virtual leaks, and hence lengthy pump-down times, small communicating passages should be made between any trapped volumes and the pumped volume. See figure 5.23.

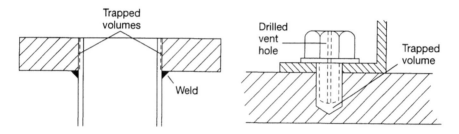

Figure 5.23 Examples of trapped volumes.

Summary

The selection of materials and components for use in any vacuum system is a very important aspect of the system design and should, of course, always be considered in close consultation with the user. However it is appreciated that the cost will always be of prime importance but undue emphasis on this aspect, such as reducing the number of flanges on the main vacuum chamber, may eventually be very expensive and time consuming. Some references have been given which apply to vacuum materials and their selection.

Acknowledgments

The author wishes to thank the United Kingdom Atomic Energy Standardisation Committee for permission to use material published in their Code of Practice AECP 100 parts 3, 5 and 11 and Pergamon Press for their permission to use material published in *Vacuum* **37** (1987).

References

Elsey R J 1975 *Vacuum* **25** 299 - 306, 347 - 61
Espe W 1966 *Materials of High Technology* Vols 1 - 3 (New York: Pergamon)
General references
Halliday B S 1987 *Vacuum* **37** 583 - 95
Laurenson L 1980 *Vacuum* **30** 275 - 81
Weston G F 1975 *Vacuum* **25** 469 - 84

6

Cleaning

B S Halliday

The main object in creating a vacuum environment for a process is to reduce, as far as is required, the presence of gases and vapours in the chamber. The cleaning process is to reduce, to an acceptable limit, the evolution or desorption of gases from surfaces during the operation of the equipment. The source of these gases can be surface soils such as swarf, oils, greases, fluxes put down during manufacture, marking inks or dyes, surface corrosion and fingerprints. Gases are also desorbed from surface layers and diffuse from the bulk of the materials. However the cleaning methods used to remove these soils should have no deleterious effect on the materials and not leave any deposits which can affect the vacuum process. Raw materials freshly machined can be easily cleaned, but assemblies of components must be cleaned as far as possible as discrete elements, as it is difficult to clean assemblies, for crevices, bolted surfaces and sliding joints may hold cleaning fluids for a considerable period; these problems must be considered at the design stage.

It is usually essential to use some form of chemical cleaning to reach the standard required and the designer should consult where possible an experienced person so that no process is chosen which could damage the materials or create problems from which recovery is likely to be both difficult and expensive. Mechanical cleaning methods should also be considered and may be adequate. It is important to understand the safety regulations for the chemical cleaning methods and manufacturers' instructions should be closely followed. It is also necessary to comply with all statutory rules of the Department of Trade and Industry (DTI) in the UK and the National Regulatory bodies in other countries. Care must be taken at all times to ensure good housekeeping; the work areas must be kept clean, tidy and well ventilated so that toxic and corrosive vapours cannot affect operators and cause damage to buildings. Cleaning fluids should be maintained in good condition and free from gross contamination which can reduce their effectiveness. Solvents, acids and other agents should be correctly stored. Personnel should be supervised at all times and be adequately trained to handle the chemicals involved and understand all relevant regulations.

6.1 GENERAL CONSIDERATIONS

6.1.1 Design considerations

Good design can enable good cleaning techniques to be employed, not only in the initial stages but also for cleaning after use. Blind holes, crevices and faulty welds must be avoided as they can become clogged with dirt during manufacture. This can cause virtual leaks and retain cleaning fluids. Blind threaded holes, close fitting joints and bolted assemblies giving trapped volumes are also some of the design features to be avoided. It is essential, at the design stage, to consider methods of assembly so that components can be individually cleaned successfully and also to permit efficient cleaning after the equipment has been in service for some time.

6.1.2 Material selection

Materials chosen should fulfil all criteria for vacuum equipment and especially the avoidance of porous materials which are not only unacceptable for their outgassing but would be absorbent to the cleaning fluids. Equipment manufactured from more than one material may be chemically incompatible with cleaning fluid and may cause deposition of one material on another and create electro-chemical cells causing corrosion, so must be cleaned individually. Materials themselves exert a vapour pressure but under normal circumstances at room temperature this is not significant.

6.1.3 Dimensions

Large equipment should be designed so that it can be cleaned in smaller parts where possible, which can be easily handled and allow correct access and removal of cleaning fluids. These smaller parts can then be re-assembled using demountable joints so that further cleaning is unnecessary. Lifting lugs must be provided, of adequate strength for lifting large components even when filled with cleaning fluid. After component part cleaning is complete, components should be temporarily packaged until final assembly is carried out and for any further handling, lint free or plastic gloves should be worn to prevent surface contamination with finger grease.

6.1.4 Surface finish

Aggressive chemical cleaning can affect dimensional accuracy and the 'roughness' can increase surface area, which must be considered where ultra-high vacuum is concerned. Ultrasonic cleaning of bearings can cause pitting which will affect their performance in service.

6.1.5 Gases from surfaces
In a leak free system the lowest pressure obtainable is largely determined by the rate gases are desorbed from surfaces and these areas should be kept to a minimum. It should be noted that at pressures below 10^{-3} mbar gas molecules are not influenced by the position of the pump as they are in molecular flow and gases evolved from one surface due to inadequate cleaning can be redeposited on another surface causing further problems.

6.1.6 Surface contaminants
Minute amounts of contamination can prevent working pressure being reached in a reasonable time. These contaminants can be cutting oils, water, paint, marking inks, finger grease, lint from cleaning rags, oxide layers saturated with water vapour and the presence of vacuum oils and greases where not intended.

6.1.7 Adsorbed gases
The amount of adsorbed gas is proportional to the surface area exposed to the vacuum and this can be increased by surface roughness including loosely adherent layers of oxide. Atmospheric gases including water vapour are adsorbed on each opening of the system and the rate of desorption governs the time of subsequent evacuation. Water vapour in particular is strongly adsorbed and desorbs slowly, but desorption time can be reduced by venting with a suitable dry gas (e.g. nitrogen).

6.1.8 Absorbed gases
In equipment operating below 10^{-7} mbar, desorbed gases can be the major pump load and must be considered. Bakeout is therefore used to obtain the required pressure more quickly by temporarily increasing the desorption rate, the evolved gases being pumped away.

6.1.9 Oxide films
Thin, tightly adherent layers which are non-porous (e.g. the colouration after argon welding) are usually acceptable, but not loosely adherent layers of porous oxide.

6.2 PROCEDURE SELECTION
This will depend on a number of factors:
(a) The required working pressure and the process for which the vacuum is required.
(b) The state of the material to be cleaned.
(c) The shape, size and weight of the components.
(d) The cleaning plant available.
(e) The safety and economic requirements.

A guide to cleaning procedure selection is shown in table 6.1, page 142.

6.2.1 The process and working pressure

The process to be carried out and the working pressure are linked and will influence the choice of procedure. Some processes, for example, may preclude the use of chlorinated solvents whereas acid treatments may leave unacceptable residues.

6.2.2 Materials

The cleaning procedure selected must not have a deleterious effect on the materials, and will depend on the state of the materials' contamination.

6.2.3 Dimensions

A large vacuum plant presents problems and may require the plant itself to be used as the solvent containment with its connecting pipework. It should be remembered that only the surfaces exposed to the vacuum require the full cleaning procedure.

6.2.4 Plant available

This may dictate the procedure selected and outside agencies may need to be contacted. Manufacturers of vacuum plant usually have satisfactory procedures of their own which are adequate. Every effort must be made to ensure that lack of in-house plant does not prevent the full required procedure being performed.

6.2.5 Safety and economics

The selection of elaborate procedures for simple requirements are uneconomic and unnecessary, e.g. it is unnecessary to use a demineralised water rinse for a diffusion pump backing line.

Toxic vapours, acids and pickles must be handled with care by trained personnel who are fully aware of the hazards and required safeguards. Manufacturers' recommendations for proprietary solvents, soak cleaners etc must be followed and procedures must fulfil all statutory requirements. All this work must be strictly supervised.

Over the last few years, environmental protection legislation, in the UK particularly, has restricted the use of some types of chemical solvent for metal cleaning. Precautions have been tightened in the use and work practice for those solvents which are permitted.

6.3 CLEANING PROCEDURES

These usually consist of a number of processes used sequentially so that a higher degree of cleanliness will require more stages of cleaning. It is again recommended that only the number of stages necessary are carried out.

6.3.1 Mechanical cleaning

General surface debris and heavy soiling can be removed by scraping, brushing and wiping. These processes should be kept to a minimum, and the use of coarse abrasives, grinding or abrasive pastes should be avoided. The surfaces should then be thoroughly washed with plenty of hot water.

6.3.2 Removal of gross contamination

After mechanical cleaning, gross contamination may be removed by a wash in hot water jets (approximately 80 °C) using an alkaline detergent. This is followed by hot water only with no detergent.

6.3.3 Removal of paints and adhesives

These may be removed by strippers or solvents compatible with the materials of the component; abrasive jet blasting may be used.

6.3.4 Removal of fluxes

Soaking and scrubbing in hot water is usually effective but the flux manufacturers should also be consulted.

6.3.5 Wet bead blasting

In this process, a jet of minute glass beads or alumina powder in water is projected at high velocity onto the surface to be cleaned. It produces a fine matt surface finish and may be used to remove oxides, paints and other surface coatings. If used for final cleaning, the plant must be reserved for this use only and a high state of cleanliness is required.

6.3.6 Dry bead blasting

Abrasive powders (e.g. alumina) can be used for the removal of surface contamination (e.g. insulator metallised coating), and a glove box is recommended to contain the powder. Both blasting processes require water rinsing to remove loose material and debris.

6.3.7 Degreasing by immersion

This is an essential stage of cleaning and may be performed more than once. The component is immersed in a bath of hot, clean, stabilised trichloroethylene (Triklone N^{TM}) for at least 15 minutes, or until the item has reached the temperature of the bath. All immersion processes are enhanced and speeded up by ultrasonic agitation of the solvent. A typical twin-tank ultrasonic liquid and vapour zone cleaner is shown in figure 6.1.

6.3.8. Vapour degreasing

This can be carried out in a special tank using stabilised trichlorethylene vapour but will only operate up to the temperature of the boiling point of the solvent (in the region of 60 °C). Thin section components may require two or three treatments. This process removes oils and greases. Freshly machined aluminium can react violently with this solvent and should therefore only be used after expert advice.

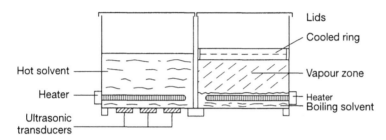

Figure 6.1 Cross-sectional diagram of a twin-tank cleaning unit.

Method of operation.

(1) After a preliminary clean to remove all loose soils, debris etc immerse in hot solvent (trichloroethylene) with ultrasonic agitation for about 15 minutes. Lift out of solvent and allow to drain dry.

(2) Immerse into hot vapour zone until vapour ceases to condense and drip off component. (NB a thin walled component will heat up quickly so that several immersions should be made.)

(3)   This is followed by a wash with hot (60 °C) water jet (no detergent).

6.3.9   Alkaline degreasing

A further stage of alkaline degreasing may be carried out (P3 Almenco P 36™), if possible with ultrasonic agitation, at a temperature of 60 °C for 5 minutes. This must be followed immediately by a further rinse using jets of demineralised water (80 °C). Continue until all particular deposits of the alkaline bath are washed away, especially from trapped areas, blind holes etc. (P3 Almenco P36 is not suitable for copper.) Alternative solvents for the immersion and vapour cleaning are:

> trichloroethane where local regulations permit
> perchloroethylene where local regulations permit
> isopropyl alcohol
> acetone.

Alternatives for alkaline degreasing:
> Almeco T516™
> Almeco P3-36™.

6.3.10 Pickling
This is a process very rarely recommended for vacuum applications. The fluids used attack the metal and oxide layer and frequently the residues are difficult to remove and are unacceptable. It produces an etched finish so that where the requirement for smooth and shiny surfaces is important it should be avoided. Pickling cannot be used when strict dimensional accuracy is required as the etching effect is non-uniform. This can be overcome by final machining after the pickling, followed by a final degrease and rinse.

6.3.11 Passivating
The corrosion resistance of aluminium and some stainless steels depends on an adherent oxide layer. When a pickling process has been carried out, it is better to re-form this layer by chemical means, than to wait for the normal oxidation process by reaction with air. A dilute solution of nitric acid will restore the oxide layer very quickly. Expert advice on this should be sought.

6.3.12 Washing
Washing is an important process as it removes the residues of aggressive fluids and other surface residues e.g. alkaline detergent degreasers. Domestic tap water can usually be used initially in a running tank or by a hose spray so that a continual supply of fresh water is in contact with the component. As tap water often contains chlorides it may not always be acceptable. The tap water rinse should be followed by demineralised or distilled water where the conductivity must not rise above 10 μS cm^{-1}. Where conductivity checks are not available, the final rinse should be by spray or rinse from newly opened bottles of distilled water. As an aid to drying, the component can be immersed in a bath of water of the correct quality at 65 °C for a time so that it will dry completely when removed.

6.3.13 Polishing
Mechanical polishing processes should be avoided as they usually cause the polishing medium to be embedded in the surface to its detriment. Chemical and electropolishing methods are usually employed for vacuum applications. Either of these processes are best carried out by specialist contractors and should be fully discussed with them before they are undertaken and samples should be made of the suitability of the process. They are expensive processes and their use is rarely justified.

6.3.14 Masking
UHV systems with all-metal seals have accurately machined flanges which can be attacked by cleaning fluids and therefore have to be masked before the process starts. This is frequently done by the use of self-adhesive tapes and it is important

that when the tapes are removed after cleaning any adhesive residue is also removed. This is usually achieved by swabbing using a lint free cloth, and suitable solvent (acetone or ethyl alcohol).

6.3.15 Drying and handling
It is essential that components should be quickly and thoroughly dried after the final wash. This can be done by use of a stream of dry, dust and oil free air, preferably hot. When drying is complete the component must be packaged or installed immediately to prevent any contamination by air-borne contaminants or by inexpert handling. Any handling should be with clean, dry and lint free or plastic gloves. It should be remembered that these gloves should only be worn when actually handling clean components and contact must be avoided with unclean surfaces, e.g. uncleaned nuts, bolts and screws, tools and the outer surfaces of components unless these too have been cleaned. The gloves must be removed when opening cupboards, drawers and touching any unclean surface. Consideration should also be given to the wearing of suitable head covering as hairs etc are detrimental in most vacuum systems.

6.3.16 Bakeout
To obtain pressures less than 10^{-7} mbar it will be necessary to accelerate the evolution of sorbed gases and residual contaminants by baking the system while continuously pumping using the system pumps.

It is necessary to bake the whole of the system including the high vacuum side of pumps, gauges, valves and any UHV pipework. This involves raising the temperature to typically 200 °C for 24 hours. The bakeout of all the UHV surfaces is necessary as any unbaked area will act as a condensing surface for desorbed gases which will then be re-distributed when bakeout ceases and the baked surfaces are cooled. Typical gases desorbed are hydrogen, oxygen, methane, carbon monoxide and water vapour. The temperature of bakeout depends on the materials of the system and the adjacent structures and will influence the materials selected e.g. if bakeout is at 300 °C elastomer seals cannot be used.

6.4 PACKAGING
It is important to consider the time the component will be in store and whether damage in transit can be expected, thus determining the type of packaging to select. In general it is recommended that all openings be covered with new aluminium foil crimped into place then followed by polyethene sheeting or pre-fabricated covers. Knife-edge flanges can be protected using scrap copper rings. It should be noted that some plastic sheeting may have the residue of release agents from the manufacturing process which may be unacceptable. The use of some adhesive

tapes which contain chlorides in their adhesive can initiate corrosion. The application of marking inks to external surfaces has also been found to initiate corrosion.

6.5 GENERAL CLEANING TECHNIQUES

Table 6.1 shows typical sequences suitable for cleaning most materials to be used for processes in various ranges of pressure. Specialist cleaning may be required and this must be decided at the time the equipment is designed.

6.6 SPECIAL CLEANING PROCESSES

6.6.1 Ceramics
(1) Preliminary cleaning can be carried out using a nylon brush and a proprietary scouring powder. Wet slurry blasting may be used. However, fully vitrified unglazed ceramic marks easily and should be handled at all stages by gloved hands.
(2) Immerse in 1 - 2% detergent solution in an ultrasonic bath for 10 minutes.
(3) Immerse in cold water bath with ultrasonic agitation for 10 minutes.
(4) Immerse in nitric acid (density 1.42), 500 ml acid with water to 1 litre in ultrasonic bath, for 5 minutes maximum (NB acid added slowly to water, wearing suitable protection).
(5) Wash in domestic water.
(6) Wash in demineralised water.
(7) Dry in a stream of hot dust and oil free air.
(8) Heat in air in oven to 1000 °C for 8 hours (NB the rates of heating and cooling should not exceed 50 °C).
If the ceramic has metal parts brazed to it then omit (4) and (8) above.

6.6.2 Glass
The cleaning of borosilicate glass (Pyrex) is usually by the traditional method of immersion in chromic acid solution but this requires stringent safety methods as it is toxic. Soak cleaners are now preferred as these are low in toxicity and residues can be removed with water washing. This method should be tested before full acceptance. See section 6.3.
Procedure for new glassware.
(1) Degrease by immersion in 5-10% solution of soak cleaner at room temperature until clean.
(2) Wash in running domestic water.
(3) Wash in demineralised water, preferably hot.
(4) Dry in stream of hot, dust and oil free air, or in an oven.

Alternative method.
(1) Degrease in trichloroethane by rinsing, swabbing or immersion.
(2) Immerse in 1-2% detergent solution in an ultrasonic bath for 10 minutes. See section 6.3.7.
(3) Wash in running domestic water.
(4) Wash in ultrasonic bath of hot demineralised water for 2 minutes.
(5) Dry.

6.7 CLEANING OF VACUUM COMPONENTS AND PLANT AFTER USE

Vacuum plant will require cleaning after it has been in service when it can no longer reach the required working pressure or when the process cannot be performed satisfactorily. However it may be part of planned maintenance which calls for cleaning during service.

Contamination may be in many forms, e.g. coating plant may require chamber walls and sight glasses cleaned, or incorrect operation may have deposited pump fluids into the working chamber.

Equipment is best cleaned if it can be broken down into separate parts so that they may all be cleaned in the manner which is required for that material, and those materials which might be affected by cleaning fluids can be treated separately eg organic sealing rings in flanges and valves. However organic seals should never be cleaned with solvents as these can easily be absorbed with later desorption problems. Special treatments may be required in special cases, e.g .the removal of metallic deposits in coating equipment.

Pumping equipment, gauges, valves etc should be cleaned and re-conditioned according to manufacturers' instructions either under in-house arrangements or by returning the pumps etc to the manufacturers or to specialist firms who recondition equipment. It should be noted that most manufacturers require details of any contamination which may be present before any work can be undertaken.

Summary
This section has outlined basic requirements only and draws attention to the need to consider cleaning processes at the design stage with tests of methods and consultation at an early stage.

Acknowledgments
The author wishes to thank the United Kingdom Atomic Energy Standardisation Committee for permission to use material published in their Code of Practice AECP 100 part 5 and Pergamon Press fo.r their permission to use material published in *Vacuum* **37** (1987). The author also thanks Dr R Reid of the Daresbury Laboratory (CLRC) for helpful discussion and reference to the Daresbury *Ultra High Vacuum Guide 1996* (The Red Book).

Table 6.1 Guide to cleaning method selection.
Note. The processes shown are the minimum required and users are reminded that the working pressure may not be the overriding requirement as the process may require a special treatment due to surface reactions during the process.

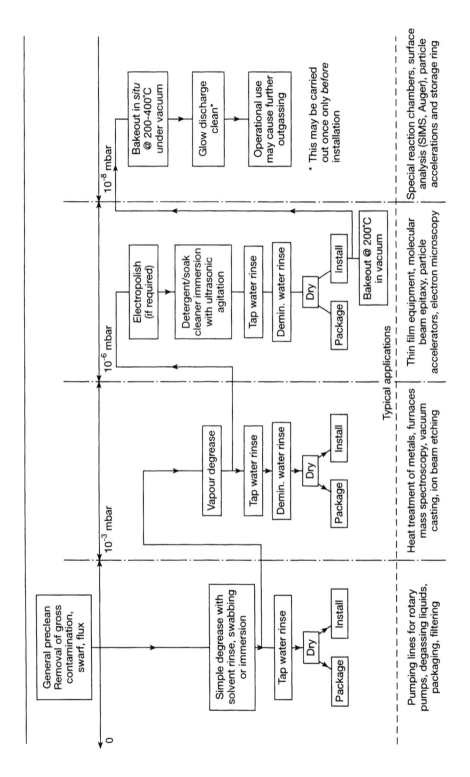

7

Leaks and Leak Detection

R K Fitch and B S Halliday

The importance of this aspect of vacuum technology must have been appreciated by one undergraduate, when answering a question in his applied physics examination, who stated that the essential requirements of a high vacuum system are a rotary pump, diffusion pump and a leak! ('Snippets' 1988). This may have been a misunderstanding but it does perhaps highlight the importance of the subject of leak detection because there is no value in designing the appropriate vacuum system and selecting the relevant materials, pumps and gauges if facilities are not available to detect and eliminate leaks. Indeed leak detection is perhaps one of the most important but sometimes one of the most tedious and frustrating aspects of vacuum technology. However much of this unnecessary effort can be avoided if a correct approach is made to the original design and construction of the vacuum system. In practice there are really two main aspects of leak detection, namely testing components and testing completed systems and any particular technique may be more or less appropriate to the one than the other. However it is first necessary to look at the nature of the leaks that can occur in all vacuum systems.

7.1 REAL AND VIRTUAL LEAKS

Leaks are normally referred to as 'real' or 'virtual', where in the former the gas passes from the external atmosphere into the vacuum chamber and the latter arises either due to the evolution of gases or vapours trapped inside the vacuum envelope in holes or channels, or due to desorption of adsorbed molecules or vapours on the inside walls and components in the vacuum system. In the latter case the most important example of this is adsorbed water vapour.

The subject of real leaks has been discussed in detail by Beavis (1970). They are normally referred to as 'pore' and 'permeation' leaks. In the case of the pore leak the external gas flows molecularly or viscously into the vacuum chamber through mechanical imperfections, or cracks in the chamber walls or holes that were previously closed but have now been exposed as a result of cleaning. The leak rate decreases with increase of temperature for viscous flow but increases with temperature for molecular flow. Furthermore both require a minimum hole size of at least about ten molecular diameters.

In contrast permeation leaks require that there is an electronic interaction between the gas and the solid and consequently leaks of this type are usually specific. The leak rate increases rapidly with increase of temperature and it also depends on the interatomic spacing in the solid. Important examples of permeation leaks are helium through glass, oxygen through silver and hydrogen and helium through neoprene and Viton O-rings.

A leak rate, Q_L can be defined as the quantity of gas which enters the vacuum space per unit time from real or virtual leaks, or both. Thus as discussed in section 2.10 we can express Q_L in the form

$$Q_L = v(dp/dt)$$

where dp/dt is the rate of rise of pressure in a closed volume, v, isolated from the pumps. If the system is being continually pumped the rate of removal of the gas is equal to the rate of entry of the gas through the leak. Then we can write

$$Q_L = S \times p_u$$

where S is the pumping speed and p_u is the ultimate pressure in the system. Hence it is meaningless to speak about a 'large leak' unless we are able to specify the volume of the system. These considerations determine whether a leak is significant or not and in a continuously pumped system a leak may be present which is small enough to be ignored. For example, in a chamber with volume 125 l and surface area 1.5×10^4 cm^2, the total surface outgassing rate (for stainless steel) would be 1.5×10^{-8} mbar l s. With a pump of speed 1000 l s^{-1} the pressure without leaks would be 1.5×10^{-11} mbar. If we now have leaks of various sizes, the resultant pressure is shown in table 7.1 below.

Table 7.1

Q_L (mbar l s^{-1})	Q_T (mbar l s^{-1})	p (mbar)
1.0×10^{-5}	1.0015×10^{-5}	1.015×10^{-8}
1.0×10^{-9}	1.6×10^{-8}	1.6×10^{-11}
1.0×10^{-10}	1.51×10^{-8}	1.51×10^{-11}

From these figures it can be seen that a small leak, well within the detection limits of all the helium leak detectors, has little effect on the final pressure. The decision has to be made whether a leak can be ignored.

It should also be emphasised that the leak rate will be independent of the chamber pressure provided that the pressure is less than about 1 mbar, because the pressure difference will be constant and equal to 1 atm. The rise of pressure with a constant real leak into a fixed volume is shown as a function of time in figure 7.1(a) and the straight line indicates that the gradient of the line, dp/dt, is constant.

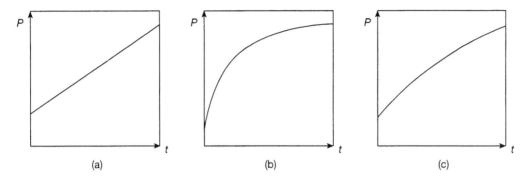

Figure 7.1 Pressure versus time curves if (a) a real leak, (b) a virtual leak or (c) both are present.

However if a similar graph is plotted for a virtual leak then the initial high rate of rise of pressure gradually falls off to zero as shown in figure 7.1(b), when the rates of adsorption and desorption of the gases or vapours from the surfaces inside the vacuum chamber are in equilibrium. The familiar case when a real leak and a virtual leak are both present is shown in figure 7.1(c). Hence the common practice of using the pressure versus time graph to identify a real leak must be used with caution. However McIlraith and Scott (1966) showed that air and vapour leaks can be distinguished by using the fact that the pumping speed of a rotary pump is much greater for a gas than a vapour. Thus if the steady pressure is p mbar before the rotary pump is isolated, and if the pressure rises at a rate dp/dt following isolation, then

$$Sp = v(dp/dt)$$

where S = speed of pump and v = the approximate volume of the system.

If the value of S is comparable with the rated speed of the pump then an air leak is present, but if a vapour is present then the measured speed will be between 100 and 1000 times smaller than S. Furthermore because the pump has a much smaller speed for a vapour it is not necessary to make allowance for the conductance of the connecting pipes between the pump and the vacuum chamber. If the presence of a vapour is suspected then it will be necessary to gas ballast the pump before taking any measurements. Virtual leaks which are due to outgassing of the surfaces and components inside the vacuum system can often be reduced to an acceptable minimum by controlled heating of the system but virtual leaks due to vapours and small trapped volumes are very difficult to find. In this case all that can be done is to establish the presence of the virtual leak and then consider any possible causes, for example due to any recent changes in design or modifications made to the system.

7.2 METHODS OF LEAK DETECTION

There are essentially two methods for the detection of leaks, namely when an excess pressure of the search gas is maintained inside (I) or outside (O) the system under test, using a number of different types of leak detector in each case.

7.2.1 Inside (I). These methods are suitable for experimental chambers and certain components in which it is possible to fill the chamber or component with a search gas at a pressure which is greater than the surrounding atmosphere. If a leak is present then the gas can be detected by an external 'probe' or 'sniffer'. In general these methods are satisfactory for applications in low, medium and high vacuum but are not sensitive enough for ultra-high vacuum systems. In addition these methods have other disadvantages due to the dangers of pressurisation and possible contamination of the vacuum system.

7.2.2 Outside (O). In this group we must include the relatively simple but important devices which must be used if we cannot obtain a pressure low enough to operate the more specialised devices. In the latter case the detector is situated in the vacuum system or in a leak detector unit which can be attached to the system or component under test, but the detector must be suitably positioned to intercept the flow of the probe gas. Furthermore the base pressure of the system or unit must be stable in order to benefit from the high sensitivity of some of these devices.

7.3 LEAK DETECTORS

The mass spectrometer is the most sensitive and most important leak detector but many small companies and institutions cannot justify the expense of one of these instruments and a number of alternative detectors will be described. By reference to the appendix at the end of this chapter it will be seen that most alternatives have considerably lower sensitivity than the helium mass spectrometer, but cost and the degree of sensitivity required may make them acceptable. In each case it will be indicated whether it is suitable for method (I) and/or (O).

7.3.1 Soap bubbles (I)
In this case the probe gas will normally be compressed air but greater sensitivities can be obtained using helium. The soap solution can be brushed on the suspect areas and bubbles will be seen if a leak is present. However it is more satisfactory if the chamber can be totally immersed in the liquid and the bubbles can be seen rising to the surface. It is possible to obtain higher sensitivities using liquids of lower surface tension.

7.3.2 Leak covering (O)

A suspected leak can be covered over with a tape or low pressure 'plasticine' such as Apiezon Q. If a leak is present then a fall of pressure will be observed. However this method should be used with care because when the tape or plasticine is removed the leak may be temporarily sealed but will almost certainly reappear at a later stage. This method should not be used for UHV applications.

7.3.3 Thermal conductivity detector (I and O)

As the thermal conductivity varies for different gases then a Pirani or a thermocouple gauge can be used as a leak detector. One of these devices is included in most vacuum systems and a leak can be indicated by an apparent change in the pressure when another gas e.g. carbon dioxide is directed over the leak. Low priced hand held monitors using the thermal conductivity principle are widely available and can also be used to detect the escape of domestic and industrial combustible gas from meters, valves, and pipeline joints. Leak detection over welds, joints, seals, etc can be carried out using helium or carbon dioxide as the trace gas. They can also be used to check leaks from gas cylinders and aerosols. Their maximum sensitivity is limited. If method (O) is being used then it will be necessary to use a fan to direct the gas from the suspect leak area towards the thermal conductivity probe. These detectors are suitable for pressures down to 10^{-3} mbar. However care must be taken as both gases are inflammable.

7.3.4 Ionisation gauge and ion pump detectors (O)

Most high vacuum systems include some type of ionisation gauge. The ion gauge current and the ion pump current are both dependent on the gas species in the system or pump and changes can be seen when a leak is present and covered with a search gas as used for the thermal conductivity gauge.

Useful safe gases for both methods are helium, carbon dioxide and argon. Acetone and isopropyl alchohol can be used with care.

7.3.5 Mass spectrometer leak detector (I and O)

The mass spectrometer leak detector is an essential piece of equipment in almost every laboratory involving vacuum equipment and the many associated technologies. This may be an already existing residual gas analyser (RGA), whether it be a magnetic sector or quadrupole type, but it does not need to have a very high resolution because it is only necessary to separate the two most commonly used gases, namely hydrogen and helium at mass numbers 2 and 4 respectively. Nevertheless the dedicated mass spectrometer leak detector is usually of the magnetic sector type because of its higher resolution at lower mass numbers and its higher sensitivity compared with the quadrupole. Its basic principles, operation and maintenance have been well documented in the *Handbook of Nondestructive Testing* edited by McMaster (1982). Helium is usually chosen as the search gas because the molecule is small and inert, it has a higher particle velocity than any other gas at any particular temperature except hydrogen, has no environmental hazards, is non-corrosive and perhaps most importantly is unlikely to be present in the vacuum system for any other reasons. In exceptional cases hydrogen may be preferred because of its higher sensitivity. In other applications it may be necessary to tune the mass spectrometer to another search gas due to the permeability of, for example, hydrogen and helium through Viton O-rings. The conventional mass spectrometer leak detector is normally a portable unit containing its own vacuum system including a rotary and a diffusion pump, liquid nitrogen trap, gauges, valves and all the associated instrumentation and electronics as shown schematically in figure 7.2. A Penning gauge is normally used because of the continual cycling to atmospheric pressure.

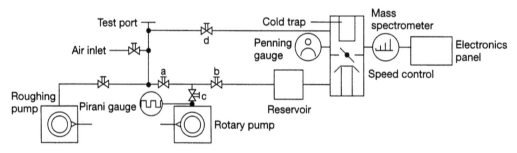

Figure 7.2 Schematic diagram of a conventional mass spectrometer leak detector.

When the vacuum system is in operation the component or system under test is connected to the 'test port' with valves b and d closed and valves a and c open. A number of helium leak detectors have a second rotary pump for roughing out the system under test while the diffusion pump is kept under working conditions by its dedicated pump. When the pressure is low enough valve a is closed and valves b and d are opened. The mass spectrometer filament is switched on when the pressure on the Penning gauge is less than ~5×10^{-4} mbar and the system under test is probed with the helium search gas. If a leak is detected this is indicated by an audio signal and/or an analogue meter on the electronics panel. The sensitivity of the detector can be increased by (i) increasing the gain on the electronic amplifier and/or (ii) reducing the effective speed of the diffusion pump by adjusting the speed control valve but still maintaining the minimum pressure to operate the filament. With this arrangement the minimum detectable leak is about 10^{-12} mbar l s^{-1}.

A different type of this leak detector has been described by Reich (1987) in which the diffusion pump has been replaced by a turbo molecular pump so that no liquid nitrogen is required and the detector is operational in less than 5 minutes whereas in the conventional design the cold trap is necessary to increase the helium concentration by condensing the vapours and thus decrease the pump-down time. Some helium leak detectors, especially the portable types, use a trapless molecular drag pump instead of a diffusion pump.

Another version of the mass spectrometer leak detector uses method (I) which incorporates a flexible probe containing a leak at one end where the sampled gases are admitted to the ion source of the mass spectrometer and with helium being the pressurising gas. The main advantage of this method compared with the conventional spectrometer is that it can test for leaks on an existing system without the need to disturb the system and the unit is usually made to be very portable. However the minimal detectable leak is only about 10^{-8} mbar l s^{-1}.

Of course it is important to be able to calibrate your leak detector at regular intervals using a standard leak containing a reservoir of helium which leaks out at a constant rate through a quartz membrane. Various leak rates can be obtained with the smallest one being about 10^{-8} mbar l s^{-1}. Jitschin *et al* (1988) have shown that these leak artefacts give a measured leak rate change of only a few per cent variation over a period of 6 years. However this type of leak is strongly temperature dependent and therefore it is desirable to conduct calibrations within a very limited temperature range. Compressed powder leaks are also used.

The four main methods of using the helium leak detector are shown schematically in figures 7.3(a), (b), (c) and (d).

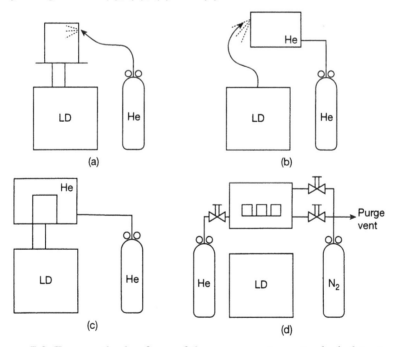

Figure 7.3 Four methods of use of the mass spectrometer leak detector.

Figure 7.3(a) represents case (O) with the external application of the search gas whereas 7.3(b) shows the over-pressure of the test vessel for case (I). In figure 7.3(c) this is a modified version of (I) in which a 'hood' encloses the whole component under test and thus it can only establish whether a leak is present or not, i.e. it does not locate the leak. Figure 7.3(d) represents the 'bombing test' for sealed units. The vessel is first pressurised with helium, then evacuated and purged with nitrogen. The leak detector then searches for helium escaping from the units under test.

A further type of detector which requires no liquid nitrogen is shown in figure 7.4 and uses a reflex or contra-flow type of leak detector.

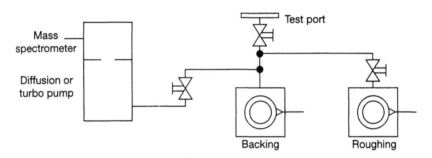

Figure 7.4 Reflex or contra-flow type of leak detector.

The mass spectrometer is kept clean and at constant pressure and the helium back-streams through a special diffusion pump where the rate is variable, controlled by the heater power. A dual port entry turbo pump or molecular drag pump can also be used. See Setina *et al* (1987).

Of course many of the new generation of leak detectors reflect the influence of modern technology and computer control. For example, many use microprocessor control with fully automatic start-up and built-in tests for such things as the device sensitivity and calibration using a built-in standard leak. In other cases there are software supported functions for fault finding in the leak detector vacuum system and many have an automatic inlet which adjusts the opening to the test port.

However the full value of the mass spectrometer leak detector can only be attained if the operator is completely familiar with the particular instrument(s) available. Even when this is the case it is essential that subsidiary pieces of equipment such as valves, connecting pipes etc are available in different sizes in order that any component or system can be quickly connected to the leak detector. In addition, polythene bags of various sizes should be available to enclose components or systems under test when a small leak is suspected, or when it is necessary to isolate one section from another. It is also important that all other vacuum systems that may need to be leak tested at some future date should incorporate a valve in the backing line so that the leak detector can be quickly connected to the system while it is still in operation.

However if a leak appears to have developed in a vacuum system it is nearly always helpful to answer these ten questions before taking any further action.

(1) Has the pressure in the system changed since yesterday?
(2) Has a different type of gas been admitted into the system?
(3) Can any sections of the vacuum system be isolated?
(4) Has the gauge calibration been checked?
(5) Are there any vapours present?
(6) Has a real leak been established even though not found?
(7) Can you tolerate the leak that has been established?
(8) Have you used some different type of cleaning material after the last exposure to the atmosphere?
(9) Have you added to or modified the interior of the vacuum system?
(10) Is there any evidence of an air leak from an RGA spectrum with peaks corresponding to masses 14, 28 and 32?

Summary

The importance of the helium mass spectrometer leak detector was again emphasised at a meeting in 1986 when it was reported that it was finding applications in many diverse areas such as automatic testing of automobile bodies, leak testing of satellites, rivets on pull rings in drink cans, large metal drums and refrigerator components, all of which are outside the conventional areas in vacuum technology. The mass spectrometer will continue to be the most important type of leak detector but there will always be an important role for the various other techniques that have been described in this chapter, particularly in establishments where vacuum technology is not a major activity. Nevertheless it should be emphasised that leak testing is somewhat of an acquired skill and thus the various 'tricks of the trade' and necessary 'green fingers' can only come with experience.

References

Beavis L C 1970 *Vacuum* **20** 233
Jitschin J, Grosse G and Wandrey D 1988 *Vacuum* **38** 883
McIlraith A H and Scott A D L 1966 *J. Sci. Instrum.* **43** 961
McMaster R C (ed) 1982 *Handbook of Nondestructive Testing* 2nd edn Vol 1 p509
  (Columbus OH: American Society for Nondestructive Testing)
Reich G 1987 *Vacuum* **37** 691
Setina J, Zavasaik R and Nemanic V 1987 *J. Vac. Sci. Technol.* A5 2650
'Snippets' 1988 *Phys. Education* **17** 10

8

Systems

B S Halliday

In this chapter seven basic vacuum systems are described by schematic diagrams and their operation discussed. The ISO and German standard graphic symbols used in vacuum technology are given on page 180.

8.1 SIMPLE ROTARY PUMPED SYSTEM

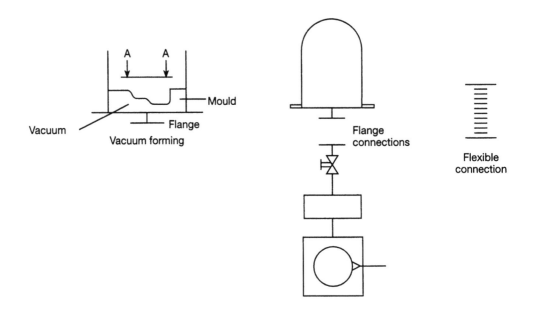

Figure 8.1 A simple rotary pumped system.

This system, using a single-stage rotary pump, would have an ultimate pressure of about 5×10^{-2} mbar. It is equipped with a flat base plate over which can be placed alternative bell jars or chambers. A filter is in series with the pumping line to filter particulate matter. Its use would be wide ranging including fluid degassing, resin mix degassing, vacuum impregnation when used with a hose to an impregnation vessel or vacuum forming when connected to a suitable mould.

8.2   DIFFUSION PUMPED SYSTEM

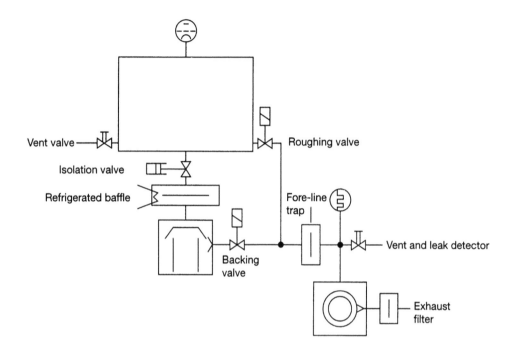

Vent valve

Isolation valve

Refrigerated baffle

Roughing valve

Fore-line trap

Backing valve

Vent and leak detector

Exhaust filter

Figure 8.2 A diffusion pump system.

This system is typical of very many vacuum rigs for a variety of applications. The work chamber is connected via an isolation valve to a refrigerated baffle or trap and then to the inlet port of the diffusion pump, the baffle or trap preventing back-streaming of diffusion pump working fluid vapour. The diffusion pump exhaust is connected through a hand operated valve (backing valve) through a fore-line trap to a two-stage rotary vane pump, this trap preventing back-streaming of rotary pump fluid vapour. The exhaust of the rotary pump enters an exhaust filter to remove vapour droplets and prevent pollution of the atmosphere. The rotary pump is also connected by a second valved line to the vacuum chamber (the roughing line). This is also protected by the fore-line trap and enables the rotary pump to be used for two functions:

(i) to maintain the required backing pressure for correct operation of the diffusion pump

(ii) to enable the chamber to be pumped out from atmospheric pressure to the starting pressure of the diffusion pump (usually about 10^{-2} mbar).

To start the system, the isolation and roughing valves are closed and the backing valve is opened. The rotary pump is started and evacuates the diffusion pump body to less than 10 mbar when its heaters can be switched on. When the pump is hot and working after about 10 minutes the backing valve is closed and the roughing valve opened to reduce the pressure in the chamber.

When the chamber pressure reaches 10^{-2} mbar the roughing valve is closed and the backing valve opened. The refrigerated baffle is now brought to working temperature. The isolation valve is opened slowly so that conditions do not exceed the critical backing pressure (section 3.2(b)) and the ionisation gauge can be switched on. A thermal conductivity gauge monitors the rotary pump pressure and an ionization gauge monitors the high vacuum pressure. If it is necessary to connect a helium leak detector it can be connected to the backing line where it can sample the full throughput of the system.

This system operates in the molecular flow region from 10^{-3} to about 10^{-7} mbar. Typical uses are: evaporation coating, crystal growing, production of TV and other electronic tubes, mass spectrometers, electron microscopes, thin film production and ion sources.

8.3 A TURBO MOLECULAR PUMPED SYSTEM

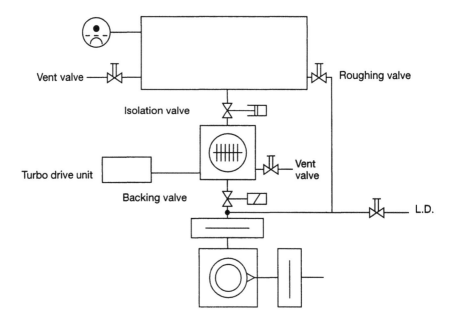

Figure 8.3 A typical turbo molecular pumped system.

In the case of a conventional turbo pump without a drag stage, a backing pump is needed, usually a two-stage trapped rotary vane pump. In the case of a compound or wide-range turbo pump with a drag stage, a dry pump of the claw, scroll, diaphragm or reciprocating piston type can be used, thus giving a totally clean and oil-free system. The turbo pump is first pumped out and switched on with the isolation and roughing valves closed. When the turbo is up to speed, the backing valve is closed and the roughing valve opened until the chamber pressure reaches 10^{-2} mbar. The roughing valve is closed, the backing valve opened first and then the isolation valve.

This system where the chamber can be isolated from the turbo pump is most used when frequent loading and unloading is necessary. In some systems where the chamber remains at its ultimate working pressure for a long period, the isolation valve is omitted and the turbo pump and backing pump are started simultaneously, because a turbo pump begins to pump a little even at low speeds and in doing so it acts as a barrier to any flow from the backing side thus maintaining cleanliness during the starting period. A dry pump would remove any possibility of contamination.

This system is clean and hydrocarbon free and is used for working pressures down to 10^{-9} mbar. Examples of its use would include particle accelerators, electron microscopes, low temperature research, surface investigations, semiconductor device production and preparation of materials for electron microscopy.

8.4 ULTRA-HIGH VACUUM SYSTEM

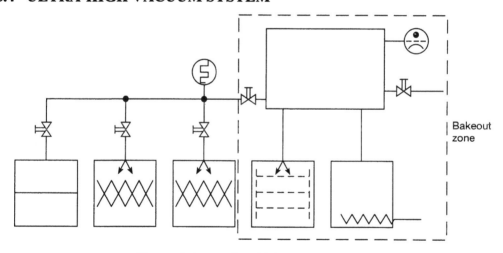

Figure 8.4 An ultra-high vacuum system.

This system is for working pressures of 10^{-8} mbar or lower and has bakeout facilities for the chamber and UHV pumps. Initial pump out of the chamber is accomplished by the use of an oil free diaphragm pump down to 50 mbar, then two adsorption pumps used in sequence down to 5×10^{-4} mbar. At this pressure the sputter ion pump can be started and when the power unit current is seen to be falling steadily the sublimation pump is switched on for a short period (about 3 minutes). When the pressure is less than 10^{-6} mbar the bakeout is started, the time and temperature depending on the ultimate pressure required. It may be necessary to use the sublimation pump occasionally to prevent the pressure rising above 10^{-5} mbar which could make the ion pumps outgas, causing the pump current to increase and the pump to overheat until the power unit trips out on overload. It should be noted that the whole of the system as shown should be baked, as any unbaked surface can condense desorbed gases during bakeout which can then travel back to previously baked surfaces when the bakeout is finished. The ultra-high vacuum gauge would be a Bayard-Alpert type ion gauge or a trigger gauge and the system would have typical applications in the semiconductor industry, molecular beam epitaxy and surface physics. Some particle accelerator storage rings with working pressures of less than 10^{-10} mbar use similar pumping techniques.

8.5 A CRYO-PUMPED SYSTEM

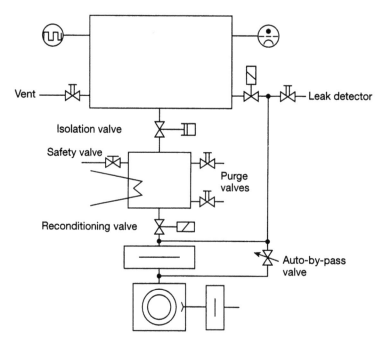

Figure 8.5 A cryo-pumped system.

A two-stage trapped rotary pump or dry pump is used to provide initial pumping, chamber roughing and pump reconditioning. The system is started by closing the isolation and roughing valve and with the reconditioning valve open, the cryo pump pressure is reduced to 10^{-1} mbar by the pump. At this pressure the reconditioning valve is closed and the helium compressor is started. The reconditioning valve is closed at this pressure as a lower pressure could cause back-streaming of rotary pump vapour and subsequent oil contamination of the cryo-pump, particularly the carbon granules.

When the indicated pump temperature reaches 20 K, the chamber is pumped to 10^{-1} mbar and the isolation valve opened. The chamber pressure must be low enough so that the admission of gas to the pump does not cause a rise in temperature of the cryo-surfaces and the desorption of gases already pumped. This admission of gas from the chamber presents a heat load through the gas by conduction from warm surfaces and by the heat of condensation of the gas being admitted. It must not be great enough to exceed the thermal capacity of the pump and the quantity of gas being pumped, rather than the pressure, is important. Typical quantities are between 30 and 600 mbar l depending on pump and chamber sizes. This corresponds to chamber starting pressures of between 10^{-1} and 3 mbar.

Reconditioning

This becomes necessary when the cryo-pump ceases to pump efficiently, generally due to the build up of solid gases causing thermal bridging between cold and warm surfaces leading to desorption of gases. The cryo-pump isolation valve is closed, the compressor switched off and the surfaces allowed to warm. Gases are liberated suddenly and uncontrollably species by species (figure 3.31) so that an efficient self-acting safety pressure relief valve is required and where necessary a ducted exhaust if toxic gases have been pumped. To reduce the conditioning time and the hazards of flammable and toxic gases, the pump can be purged with room temperature dry nitrogen and some pumps have purge valves fitted for this purpose. When the pump surfaces have reached room temperature, the cryo-pump can be pumped again before restarting the compressor by the rotary pump using gas ballast to remove any condensable vapours successfully. No hot gauges should be open to the cryo-pump during regeneration, to prevent explosions. This type of system can be used for many laboratory applications, for thin film production, ion implantation and evaporative coating, if the processes need to be completely hydrocarbon free.

8.6 A LARGE FULLY AUTOMATED MULTI-PUMPED SYSTEM

The system, shown in figure 8.6, has four diffusion pumped systems. RP1 can be isolated from its diffusion pump which can then be backed by RP2. RP1 now acts as the backing pump for the Roots pump for fast roughing of the chamber and a leak detector port is available for testing the main chamber. Pressure monitoring is by thermal conductivity gauges, Penning gauges and a quadrupole mass spectrometer for residual gas analysis. The main chamber would operate at about 10^{-7} mbar. The control and monitoring would be by computer.

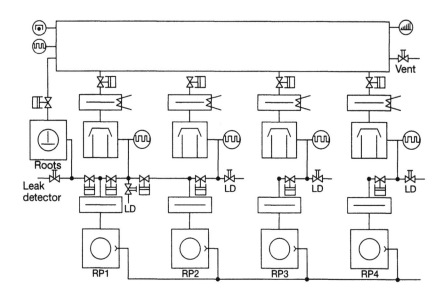

Figure 8.6 A large multi-pumped system.

8.7 A PARTICLE ACCELERATOR PUMPING SYSTEM

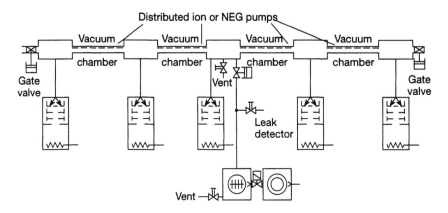

Figure 8.7 A large cyclic particle accelerator vacuum system.

This system could be used in a synchrotron where the vacuum chamber has a small cross-section (i.e. 200 mm × 100 mm) but is of great length, of the order of hundreds of metres. An example is the electron synchrotron at CERN, Geneva which is 27 km in circumference and operates at a mean pressure of 3×10^{-12} mbar. Access to the chamber is restricted by long steering magnets so that pumping stations would be spaced at about 5 m distances. The annular chamber is divided into a number of convenient sectors containing about four to six magnets, and gate valves are used to separate these sectors for individual sector testing, commissioning and maintenance. Each length is pumped by sputter ion pumps and sublimation pumps. In some accelerators distributed ion pumps or non-evaporable getter pumps are used. Roughing is provided by valved turbo molecular pumps and the whole system is remotely controlled by computer. Pressure measurement is by thermal conductivity gauges, Penning gauges and the ion pump currents, with residual gas analysers, again using computer displays.

Note
Specification and evaluation of vacuum system purchases have been discussed in a paper by O'Hanlon and Bridwell (1989).

Summary
Seven basic types of vacuum system have been described as examples but many different versions of these are in use. Details and information of these can be often found in manufacturers' literature or in specialist review articles in the various vacuum journals.

Reference
O'Hanlon J F and Bridwell M 1989 *J. Vac. Sci. Technol.* A7 202

Appendix A   Maximum evaporation rate from a surface

Consider the surface of a material of molar weight M maintained at temperature T. Suppose that the material has vapour pressure p_e at this temperature. In circumstances in which the molecules evaporating from the surface are directly pumped away so that there is no returning flux of molecules back onto it, the equilibrium situation depicted in figure 1.8 cannot be established, and the number of molecules evaporating per second per unit area of surface is given by equation (1.6) as

$$J_E = \frac{p_e N_A}{\sqrt{2\pi MRT}}$$

The mass flow flow rate is just $m \times J_E$ where m is the mass of a molecule. Therefore

$$\text{mass flow rate} = \frac{p_e N_A m}{\sqrt{2\pi MRT}} = p_e \sqrt{\frac{M}{2\pi RT}} = \frac{p_e}{7.2}\sqrt{\frac{M}{T}} \quad \text{kg m}^{-2}\text{ s}^{-1}$$

where p_e is in Pa and M in kg.

This is the maximum possible evaporation rate. The evaporating flux will be distributed in direction according to the Knudsen cosine law which is discussed in appendix D.

Appendix B Molecular drag

B.1 Molecular drag along a channel closed by a wall which moves at high velocity

Consider a long rectangular channel of width w and height h with coordinate x along its length as shown in figure B1. Its open side is effectively closed by a plate which moves continuosly with high velocity U in the x direction. Assume that gas in the channel is in the molecular state so that its mean free path λ is greater than h.

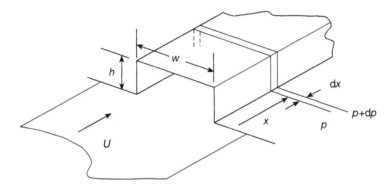

Figure B1 Channel geometry.

The motion of the high speed plate will drag molecules in the x direction by the mechanism described in section 1.15 of the text. The tendency therefore is for the number density of molecules, and hence the pressure, to increase in the x direction. However the pressure gradient thus created will act against the direction of drag. As the molecules hit the stationary walls of the channel they will lose their acquired momentum. Newton's second law of motion - 'force = rate of change of momentum' must apply. Therefore, applying it to the gas between x and $x+dx$, bounded by planes to its left and right at pressures p and $p+dp$ respectively, between which gas molecules have density n

$$hw\,dp = Jw\,dx\,mU$$

in which J is the impingement rate of molecules. Substituting for J from equation (1.8) and rearranging gives

$$\frac{dp}{p} = \frac{U}{h}\sqrt{\frac{M}{2\pi RT}}\,dx$$

Whence, by integration,

$$p/p_0 = \exp\left\{\frac{U}{\bar{v}}\sqrt{\frac{2}{\pi}}\frac{x}{h}\right\}$$

where $\bar{v}$ is the mean molecular velocity given by equation (1.2). This result is a theoretical zero-flow compression ratio. Values achieved in practice are much smaller.

B.2 Slowing of a levitated rotating sphere due to molecular drag (SRG)

Consider as in figure B2 a sphere of mass M and radius R levitated and spinning with angular velocity ω about the axis shown. Its moment of inertia for this motion is $I = (2/5)MR^2$. It is located in gas at pressure p whose molecules have number density n and bombard it at an impingement rate J.

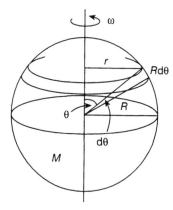

Figure B2 Geometry for analysing deceleration of sphere.

The <u>average</u> direction of arrival of molecules from the gas onto the surface of the sphere at any part of its surface is perpendicular to it. Let us assume that the molecules stay on the surface, if only briefly. In acquiring the circulating motion of the surface at any point the molecules, by Newton's third law, exert a force on the surface which is opposite to the one that gives them this new motion. This causes a slowing torque about the axis of rotation. In desorbing at a later time, the average direction of departure from the surface is perpendicular to it in the rotating reference frame of the sphere, so there is no average torque associated with departure - all the influence on the motion of the sphere takes place at the time of molecular arrivals.

Consider molecular impacts on an elemental ring at angle θ which has radius $r = R\sin\theta$ and presents an area $2\pi rR\,d\theta$ to the gas. Arriving molecules of mass m are given a sideways velocity $r\omega = R\sin\theta\omega$ in a direction which is transverse to the perpendicular direction of arrival and therefore the momentum change per second, and hence the force on this element of area is

$$J 2\pi rR \, d\theta \, m \, r\omega$$

Multiplying by a further r to give the torque about the axis and substituting for r leads to

$$\text{torque} = 2\pi \, m \, J\omega \, R^4 \sin^3\theta \, d\theta$$

Integrating this from $\theta = 0$ to π gives the total torque as $(8\pi/3)\omega J R^4 m$.

Again applying Newton's second law, in the form 'torque = rate of change of angular momentum' gives

$$\frac{2}{5} MR^2 \, \dot{\omega} = \frac{8\pi}{3} \omega J m R^4$$

whence

$$\frac{\dot{\omega}}{\omega} = \frac{20}{3} \frac{\pi \, J m R^2}{M}$$

Substituting for J from equation (1.7), expressing the mass M of the sphere in terms of its density ρ and diameter $d = 2R$ and simplifying gives the fractional deceleration rate as

$$\frac{\dot{\omega}}{\omega} = \frac{10}{\rho d} \sqrt{\frac{M}{2\pi RT}} \, p$$

So finally

$$p = \frac{\rho d}{10} \sqrt{\frac{2\pi RT}{M}} \left(\frac{\dot{\omega}}{\omega} \right)$$

Thus is pressure p related to the slowing rate of the sphere's rotation. If momentum exchange between molecule and surface is incomplete, then the slowing is less and this is accounted for by the introduction of the coefficient σ in the denominator of the above equation as discussed in section 4.6.

Appendix C   Reynolds' number expressed in terms of throughput Q

For a pipe of diameter D carrying gas of density ρ and viscosity coefficient η which travels with velocity u the dimensionless Reynolds' number is defined as

$$Re = \frac{\rho u D}{\eta}$$

The gas density ρ may be expressed as the molar mass M divided by the molar volume $V = RT/p$. Hence

$$\rho = \frac{M}{V} = \frac{Mp}{RT}$$

The volumetric flow rate $\dot{V}$ is related to the flow velocity and cross-sectional area of the pipe by the equation

$$\dot{V} = \frac{\pi D^2}{4} u$$

Substituting for ρ and u in the expression for Re gives

$$Re = \frac{4M}{\pi RT\eta}\left(\frac{Q}{D}\right)$$

For air with molar mass 0.029 kg η is 18.3×10^{-6} N s m^{-2} at 295 K and inserting these values into the equation gives

$$Re = 0.82\left(\frac{Q}{D}\right) \quad \text{for } Q \text{ in Pa m}^3 \text{ s}^{-1} \text{ and } D \text{ in m.}$$

Re-expressed in mbar, litre, cm and seconds this becomes

$$Re = 8.2\left(\frac{Q}{D}\right) \quad \text{for } Q \text{ in mbar l s}^{-1} \text{ and } D \text{ in cm.}$$

Flow is turbulent if the dimensionless quantity Re is greater than 2000, that is, if

$$8.2\left(\frac{Q}{D}\right) > 2000 \quad \text{or} \quad \left(\frac{Q}{D}\right) > 244$$

which is the required result.

Appendix D   The Knudsen cosine law

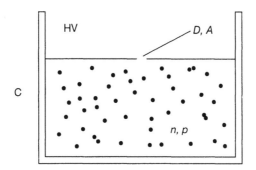

Figure D1   To illustrate the cosine law.

Figure D1 represents an enclosure C containing gas or vapour at pressure p and associated number density n. It is complete except for a small hole of area A and diameter D in a thin part of its wall, through which molecules escape into a surrounding high vacuum HV. Suppose that p is sufficiently small that the mean free path λ of the molecules inside the enclosure is at least ten times larger than the diameter of the hole through which they escape. The gas in the vicinity of the hole is therefore in a molecular state. There will be a negligible amount of molecule - molecule scattering in this region, and molecules in the bulk of the gas which are heading towards it will pass straight through to the high vacuum on the other side. This kind of passage of molecules through an aperture is called effusion, and contrasts with the situation at higher pressures in which gas streams through it as a fluid. If a mechanism exists to replace the molecules lost and maintain the number of molecules within constant, there is a steady effusive flow out of the enclosure. A vapour in equilibrium with a parent liquid or molten phase within the enclosure can be thus maintained by supplying the necessary latent heat of evaporation, as happens by design in Knudsen cells.

Under the conditions described above, because the gas or vapour is essentially in equilibrium and without bulk movement in any direction, all directions of individual molecular motion are equally probable. The flux of molecules escaping through the aperture in a direction perpendicular to the plate, which constitutes the flux which leaves the enclosure in the normal direction, consists of those which were travelling towards the hole in the same direction inside the enclosure. Their number is proportional to the area A of the hole. In a similar way, the flux emerging at an angle θ to the normal is determined by the size of the aperture for the oncoming gas flux travelling in this direction inside the enclosure, which is now $A \cos \theta$. The flux in any direction within the enclosure is the same, but the 'escape window' appears progressively smaller as the direction becomes more oblique. Figure D2(a) shows the essential geometry of the situation; the plate is assumed to be of negligible thickness.

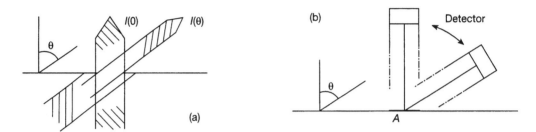

Figure D2 (a) Effusion geometry; (b) scattering geometry.

Thus the flux $I(\theta)$ emerging through the hole at angle θ may be expressed as

$$I(\theta) = I(0) \cos \theta$$

Fully defined, this is the number of molecules per second from an aperture of area A travelling into unit solid angle at angle θ to the normal. The expression above is the statement of Knudsen's cosine law of effusion. Determining the total flux I_T involves the element of solid angle $2\pi \sin \theta \, d\theta$ at angle θ and integrating the expression $I(\theta) \times 2\pi \sin\theta \, d\theta$ over the total solid angle (2π) available. I_T is thus $J \times A$, with J given by equation (1.6). It may be shown also that $I(0) = I_T / \pi$. The ideal effusion profile then has the form shown in figure 2.11. In practice, because wall thicknesses are finite the profiles are modified, as described in chapter 8 of the text by Hudson referenced in chapter 2.

Consider now the arrival of gas molecules at a surface and their scattering from it back to the gas. When, as is usually the case, the interaction of the molecules with the surface results in their leaving it in a random direction, the distribution of the scattered molecules follows the cosine law. This may be understood using similar reasoning to that used above. It is essentially to do with the obliqueness factor. Because of the isotropic nature of gas in equilibrium a detector measuring the flux from this locality would record at any angle the same value as that measured in the normal direction, $I(0)$. However, as shown in figure D2(b), at angle θ it would be registering the arrival of flux from a larger area, namely $A \div \cos \theta$, where A is the area seen by the detector pointing perpendicularly at the surface. Therefore, the amount of gas scattered at this angle which comes from the area A is the $I(0)$ value reduced by the factor $\cos \theta$. Hence, in the direction θ, $I(\theta) = I(0) \cos \theta$ and this is the Knudsen cosine law of molecular scattering. In optics, where it is referred to as

Lambert's law, a related phenomenon is the equal brightness, irrespective of the angle of view, of illuminated matt surfaces.

Evaporation from surfaces has been verified to follow the cosine law in many cases and relevant information will be found in the article by R Glang (chapter 1 of the *Handbook of Thin Film Technology*, edited by L I Maissel and R Glang, New York: McGraw-Hill, 1970).

The above is a description of scattering phenomena as they find application in vacuum practice, on surfaces which are technically rough and likely in many circumstances to be covered by adsorbed gases. But the reader should be aware that the subject has complexities. The status of the cosine law itself has been the subject of long standing discussions and the refinements of experimental techniques in surface science which probe molecule - surface interactions reveal increasing detail of basic scattering processes. The article by G Comsa (1994 *Surface Science*, **299/300** 77 - 91) is a good introduction to these matters.

Note added in proof: The discussion of the matters for which the reader has been referred to the article by R Glang in an earlier paragraph is updated in the contribution by B B Dayton in *Foundations of Vacuum Science and Technology* cited in Further Reading.

Appendix E   A derivation of the Knudsen formula for molecular flow through a pipe (equation (2.20))

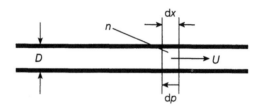

Figure E1 Pipe geometry for molecular flow calculation.

In figure E1 consider a short length dx of pipe of diameter D across which there is a pressure difference dp. Let the number density of molecules in this element dx be n and suppose an average drift velocity u can be assigned to their motion. The throughput at the element is

$$Q = kT(\mathrm{d}N/\mathrm{d}t) = kT(\pi D^2/4)nu \tag{E1}$$

By Newton's second law, equating the force across the element to the rate of transfer of momentum to the pipe wall gives

$$(\pi D^2/4)\mathrm{d}p = (\pi D\ \mathrm{d}x)(J)mu \tag{E2}$$

where the bracketed terms on the right-hand side of the equation give the number of impacts on the interior area of the wall which destroy an average drift momentum mu for each molecule. Substituting for J from equation (1.6) the above equation yields

$$nu = (D/m\bar{v})(\mathrm{d}p/\mathrm{d}x)$$

whence, substituting $\bar{v} = \sqrt{8kT/\pi m}$ from equation (1.1) and using equation (E1) gives

$$Q = \frac{\pi D^3}{16}\sqrt{\frac{2\pi kT}{m}}\ \frac{\mathrm{d}p}{\mathrm{d}x}$$

so that for a long pipe of length L across which the pressure drop is p_1 - p_2

$$Q = \frac{\pi D^3}{16}\sqrt{\frac{2\pi kT}{m}}\ \frac{p_1 - p_2}{L}$$

whence

$$C = \frac{\pi D^3}{16} \sqrt{\frac{2\pi kT}{m}} = \frac{\pi D^3}{16} \sqrt{\frac{2\pi RT}{M}}$$

This treatment is over-simplified in its assumption about drift velocity and the factor $\pi/16$ should be replaced by $1/6$ (see Loeb 1961) to give, correctly,

$$C_L = \frac{D^3}{6L} \sqrt{\frac{2\pi RT}{M}}$$

Reference

Loeb L B 1961 *The Kinetic Theory of Gases* 3rd edn (New York: Dover)

Appendix F   Analysis of a simple system

Figure F1 depicts in purely schematic form, with no valves or gauges, the gas flow paths in a system consisting of a chamber of volume v connected via a short pipe of conductance C to a diffusion pump D backed by a rotary pump R. The paths from the chamber are either through C to D and thence R or directly to R by a roughing line of conductance C_R.

The chamber is made of stainless steel and has a volume of 25 1 and internal surface area 0.5 m². The pipe with conductance C is 14 cm in diameter and 15 cm long. The pump D is well baffled with inlet diameter 14 cm, a speed of 300 1 s⁻¹ at 10^{-5} mbar, and when blanked off the ultimate pressure of the pump is 3×10^{-7} mbar. The rotary pump speed is 24 m³ h⁻¹ = 6.6 1 s⁻¹.

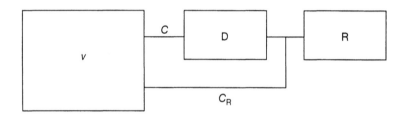

Figure F1 Schematic diagram of a simple system.

Using equation (2.24) the conductance of the 15 cm pipe is 1010 1 s⁻¹. Therefore by equation (2.13) the speed S^* at the chamber is 300×(1010/1310) = 232 1 s⁻¹ at 10^{-5} mbar. After one hour's pumping from atmospheric pressure the system pressure is 3×10^{-6} mbar; after 10 hours it is 8×10^{-7} mbar and after 50 hours 4×10^{-7} mbar, which is the lowest pressure attained. By equation (2.29) the throughput at this lowest pressure is $(4 \times 10^{-7})(232) = 0.93 \times 10^{-4}$ mbar 1 s⁻¹. This corresponds to a specific gassing rate of $(0.93 \times 10^{-4})/5000 = 2.35 \times 10^{-8}$ mbar 1 s⁻¹ per cm². This is about twice the 'typical' value mentioned in section 2.4, though since the pressure achieved is only twice the ultimate pressure of the pump when blanked off, the pump's speed may be less than 232 1 s⁻¹ so that the throughput is overestimated.

The rotary pump has a speed of 6.6 1 s⁻¹ and an ultimate pressure of 8×10^{-4} mbar. By equation (2.14) the roughing line, which is 4 cm in diameter and 1 m long, has conductances C_R of 350 and 35 1 s⁻¹ at mean pressures of 1 and 0.1 mbar respectively. For pressures down to 1 mbar $C \gg S$ and one may take a time constant $(v/S) = 25/6.6 = 3.8$ s for pumping. According to equation (2.31) the time to rough down from 1000 to 1 mbar is about 25 s, which is approximately what is observed. It takes a further 40 s to reach 10^{-2} mbar, but at this stage the pipe conductance has fallen to an order less than the speed of the pump, so that the speed at the chamber is, by equation (2.13), substantially reduced. In changing from roughing to backing at a diffusion pump inlet pressure of 5×10^{-2} mbar the rotary pump holds the fore-line pressure amply below the critical backing pressure.

Appendix G   Systems with distributed volume

A simple but instructive example is shown in figure G1, in which a long pipe of length L and diameter D is closed at one end and attached to a pump of speed S at the other. This example is analysed fully by Lewin (1965), and the important result for the steady state pressure distribution, assuming molecular flow conditions, quoted here.

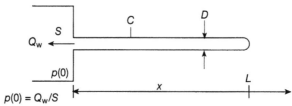

Figure G1 System with distributed volume.

Let q_G be the specific gassing rate. Since $L \gg D$ the total gas load from the interior wall is $Q_w = \pi D L q_G$ and this emerges from the open end into the pump. The analysis naturally subdivides the conductance but the conductance C of the pipe as a whole (=12.4 D^3/L for nitrogen) features in the final expression for pressure $p(x)$ as a function of distance x from the end attached to the pump. The result is

$$p(x) = Q_w/S + (Q_w/CL^2)(Lx - x^2/2)$$

where $p(0) = Q_w/S$ is the pressure at the open end of the pipe. This is a parabolic dependence of pressure on distance as shown in the second figure below. It may be further deduced that the pressure drop across the pipe is $p(L) - p(0) = Q_w/2C$ and that the average pressure is $\bar{p} = p(0) + Q_w/3C$.

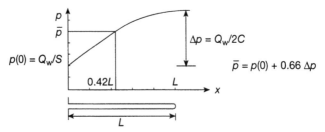

Figure G2 Pressure distribution through the distributed volume.

Clearly this result would also be of value in analysing symmetrical or repeated structures.

Reference

Lewin G 1965 *Fundamentals of Vacuum Science and Technology* (New York: McGraw-Hill)

Appendix H Specifying and measuring flow

Flow may be specified in a number of ways. They are related in the set of statements:

$$Q = p\,\dot{V} = \dot{n}_M\,RT = \left(\frac{\dot{W}}{M}\right) RT$$

which are discussed in section 2.5.2 of the text. Q is the throughput, $\dot{V}$ the volumetric flow rate at pressure p, $\dot{n}_M$ the number of moles flowing per second and $\dot{W}$ the mass flow rate in kg per second of gas of molar weight M (M is expressed in kg as discussed in section 1.4). R is the universal gas constant and T the absolute temperature.

The quantities $\dot{W}$ and $\dot{W}/M = \dot{n}_M$ are clearly mass flow rates referred to kg and molar bases respectively. If, as is frequently the case, the temperature T is constant and specified (as, say, 22° C = 295 K), then as can be seen from the equations above, the throughput $Q = p\,\dot{V}$ is directly proportional to the mass flow rate. The throughput Q is easily assembled and manipulated, and this accounts for its usefulness in describing flow. For example, down pipes which have negligible gassing off their walls, it is a conserved quantity; $Q_{in} = Q_{out}$.

Other measures of mass flow rate which are important and in widespread use derive from the definition of the *standard litre* and the similarly defined *standard cc*. They are the *standard litre per minute* and the *standard cc per minute,* **slm** and **sccm** respectively (1 cc is shorthand for 1 cm^3 , a cubic centimetre). These are mass flow rates which describe the quantity flowing per minute if that quantity were under STP conditions of 1013 mbar and 0 °C = 273 K. The identity of the gas and its molar weight must be known for the specification to be meaningful, and so if a measuring instrument records flow in either of these units it will be for a specific gas. A mass flow rate of 1 sccm of nitrogen ($M = 0.028$ kg) is seven times the flowing mass of 1 sccm of helium ($M = 0.004$ kg). Also, as noted in the text, p 15, a flow rate of 1 slm (or sccm) in conditions other than standard must be carefully interpreted. Thus a flow of 1 slm at 273 K and a non-standard pressure of 506.5 mbar will be a volumetric flow rate of 2 litres per minute. If, further, the temperature were 30 °C = 303 K then the volumetric rate would be 2×(303/273) = 2.22 litres per minute. In essence equation (1.10) of p 15 applies, modified so that volumes V become volumetric flow rates $\dot{V}$. Conversely, a volumetric flow rate of 2.22 litre per minute at 30°C and 506.5 mbar when expressed in slm is

$$2.22 \times \frac{273}{303} \times \frac{506.5}{1013} = 1 \text{ slm}$$

The opening statement of equations above also shows that in a steady state of flow at constant mass flow rate, if temperature <u>does</u> vary along the flow pathway, then it is the quantity Q/T which is conserved. Thus for a process which has a throughput of 1 mbar l s^{-1} at 295 K, the throughput in a region where the local temperature is $2\times295 = 590$ K will be 2 mbar l s^{-1}.

The practical methods of measuring flow are surveyed in chapter 6 of O'Hanlon's text referenced at the end of chapter 2. Depending on the range of flow involved, they exploit mechanical effects of flowing gas, the conductance equation (2.9), $Q = C \times (p_1 - p_2)$, in various ways or thermal effects.

In mechanically based devices called 'gap meters' (also known as 'rotameters') the flow to be measured (at atmospheric pressure) is passed vertically upwards through a conical tube of widening bore in which a light suitably shaped float of a given weight rests at equilibrium in the gas stream at a height determined by the flow rate. These devices directly measure the volumetric flow rate of a gas, and calibration is specific to a particular gas, because the force which supports the float stationary in a given flow depends on the gas density. Flow rates down to about 10 cm^3 per minute are measurable.

Throughput measurements exploiting the conductance equation are relatively straightforward in molecular flow, for which the conductance value of a simply shaped aperture can be accurately estimated. By measuring the upstream and downstream pressures the throughput is determined. Proper random entry conditions at the aperture must be assured and pressure measurement at the higher pressure side is ideally by a spinning rotor gauge, whose high accuracy becomes directly incorporated into the throughput determination. If, as is sometimes the case, the downstream side conditions are at very high vacuum, and at a pressure two orders or more less than those at the other side of the aperture, then the downstream side pressure is insignificant and with sufficient accuracy, $Q = C \times p_1$.

The measurement of mass flow rate by thermal methods is based on the heat transfer to the flowing gas from the heated wall of a pipe through which it flows. The flow-dependent property exploited is the displacement in the direction of flow of the resulting temperature profile compared with the zero flow case. Laminar flow conditions are established in a thin-walled stainless steel tube of bore less than 1 mm typically, overwound with high resistance heater windings made from material which has a marked variation of resistance with temperature. Heat sources can thus also be used as temperature sensors. Various methods for determining mass flow rate are possible. With three windings, a central heater and two outer ones to serve as temperature sensors, the difference between the higher downstream temperature and the upstream value may be measured by a bridge circuit. The heat energy input rate to the gas is the molar flow rate × the molar specific heat × the temperature difference. The power input and bridge out-of-balance signals are processed to give the flow rate. In another scheme of operation, the power needed to maintain a known temperature profile along the heated tube, which is proportional to the mass flow rate, is the basis of the measurement. In all cases instruments are calibrated for a particular gas, typically nitrogen, and correction procedures have to be used when interpreting their readings for other gases. Flows down to 1 sccm or less may be measured and depending on their magnitude, large flows may be divided within a device if necessary so that a known fraction goes through the sensor while a majority bypasses it. Frequently the devices incorporate facilities to control a downstream valve which regulates the flow.

Thermal flowmeters are extremely sophisticated instruments in both their sensor design and control/processing electronics. The need for increasing accuracy in their performance as demands become more stringent, especially in the semiconductor industry, means that they are in a continuing state of development and investigation. The article by L D Hinkle and C F Mariano (1991, *J. Vac. Sci. Technol.* A9 2043) discusses their fundamental action and, more recently, S A Tyson (1996, *J. Vac. Sci. and Technol.* A14 2582) has presented a critical evaluation of their performance.

Appendix I   The maintenance of vacuum equipment (B S Halliday)

The well known phrase 'If it ain't broke don't fix it' is very appropriate to the operation of vacuum equipment but should not be taken to mean that recommended maintenence schedules can be ignored and actions postponed if things seem to be going well.

All maintenance depends on the length of running time and the type of usage, in a similar way to car maintenance. The importance of keeping good records is paramount so that basic system parameters are known. They serve as checks when faults do occur, and there is a 'hard' record to refer to.

Records should include:

(1) Running time (a) with plant working.
                   (b) with plant idle.
(2) All measurable pressures (a) with plant working.
                              (b) with plant idle.
(3) If possible a residual gas analysis (a) with plant working.
                                         (b) with plant idle - keep for reference.

These records should be updated at regular intervals to show any gradual fall-off in performance, and then corrective action taken to prevent complete breakdown.

Detailed maintenance recommendations

Pumps.
(a) Rotary.   Check fluid level as recommended by maker
                Check noise level
                Check electrical connections
                Check ultimate pressure - using external gauge if necessary
                Check condition of zeolite in fore-line trap, if fitted.
(b) Diffusion. Check fluid level as recommended
                Check electrical connections
                Check base pressure if an isolation valve and pump gauge fitted
                Check general outward appearance for overheating etc
                Check cooling connections or fan if fitted.
(c) Turbo.   Check for unusual noise, vibration
                Check run-up time
                Check electrical connections and correct sequence by operation controller

Check shut-off pressure if possible

Check cooling connections or fan, if fitted.

(d) Ion.   Check start-up time (refer to records). If excessively long and a high initial pressure rise occurs, consider complete reconditioning or replacement.

Check H.V. connection for arcing

Check leakage current using pump power supply, and with pump at atmospheric pressure.

(e) Cryo.   Carry out inspection according to maker's manual

Check He pressures in external compressor

Check compressor absorbers for hours already run - see manual

Listen to displacer for strange noises, clicks, squeaks etc

Check temperature on low temperature stage, and if deteriorating, consider reconditioning, i.e. warming up and purging pump.

Valves.   Check operation, mechanical, electrical, or pneumatic.

Gauges.   Record readings on a day-to-day basis - compare with new condition. Look for slow changes - this could be pumps or the gauge itself

Recondition according to maker's manual

Look for electrical leakage in Penning gauges due to insulator metallisation.

Summary.   Keep good records at frequent and regular intervals

Investigate any reported irregularities by users

Establish a regular maintenance inspection

Record results of leak tests to be aware of any repeated similar faults.

Further Reading

O'Hanlon J F 1989 *A User's Guide to Vacuum Technology* (New York: Wiley)
An arguably indispensable text for the active vacuum user. Comprehensive in scope, with good discussion of underlying principles, descriptions of equipment and operational concerns, it includes many useful data tabulations and appendices. Particularly useful in residual gas analysis.

Wutz M, Adam H and Walcher W (eds) 1989 *Theory and Practice of Vacuum Technology* (Braunschweig: Vieweg, marketed in the UK and USA by Wiley)
Translated into English by W Steckelmacher, this is a comprehensive survey of the subject, thoroughgoing in its treatment and an excellent resource. A distinctive feature of considerable value is the inclusion of worked examples throughout the text.

Roth A 1990 *Vacuum Technology* 3rd edn (Amsterdam: Elsevier Science)
A broad coverage of the subject with substantial early chapters, comprising about a third of the book, which deal with the physics and physical chemistry of the underlying phenonema.

Delchar T A 1993 *Vacuum Physics and Techniques* (London: Chapman and Hall)
A text in the publisher's series *Physics and its Applications*. An ideal text for undergraduate physicists and others coming to the subject with a background in physics or chemistry.

Harris N S 1989 *Modern Vacuum Practice* (London: McGraw-Hill)
There is probably no better general introduction to the subject from the practical point of view. The treatment is largerly non-mathematical and the author's experience in teaching the principles and practice of the subject to those directly involved in using or maintaining vacuum equipment is evident throughout. A very good entry point to the subject for readers from many backgrounds and a valuable work of reference on practical matters.

Bigelow W C 1994 *Vacuum Methods in Electron Microscopy* (London: Portland)
A very broad range of vacuum techniques is employed in electron microscopy, so that despite the apparently specialist audience indicated in the title, to whom it is probably indispensable, this is a text with a wider appeal. The underlying principles of the subject and their application are clearly explained in a thorough and accessible way, and there is much sound advice on practical matters throughout.

Lafferty J M (ed) 1998 *Foundations of Vacuum Science and Technology* (New York: Wiley)
  This completely new version of a standard work, previously published under the title *Scientific Foundations of Vacuum Technique* promises to be as valuable in scope and authority as its predecessors.

Redhead P A, Hobson J P and Kornelsen E V 1968 *The Physical Basis of Ultrahigh Vacuum* (London: Chapman and Hall), reprinted in the American Vacuum Society's series of classic texts (AVS: Woodbury OH 1993).
  Invaluable for its description of fundamental principles and their application, and still highly relevant.

Weston G F 1985 *Ultrahigh Vacuum Practice* (London: Butterworths)
  A text surveying the principles and practice of attaining and measuring ultra high vacuum.

Welsh K M 1990 *Capture Pumping Technology: an Introduction* (Oxford: Pergamon)
  Emphasising practical aspects of its subject matter, and presenting the underlying fundamentals in a very accessible way, this is a most informative and valuable book.

In the English language the *Journal of Vacuum Science and Technology*, which is the official journal of the American Vacuum Society, published by the American Institute of Physics, and the journal *Vacuum* published by Elsevier Science are primary resources. They report new developments in the subject and the proceedings of conferences which are held on the wide variety of topics embraced by the technology.

Standard graphic symbols

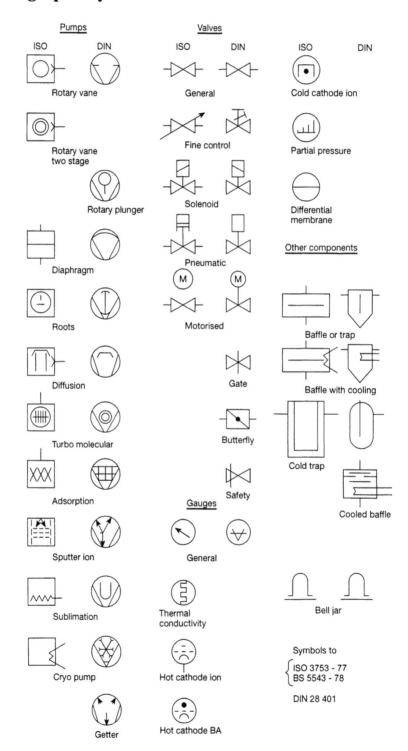

Pumps

ISO	DIN
Rotary vane	
Rotary vane two stage	
	Rotary plunger
Diaphragm	
Roots	
Diffusion	
Turbo molecular	
Adsorption	
Sputter ion	
Sublimation	
Cryo pump	
	Getter

Valves

ISO	DIN
General	
Fine control	
Solenoid	
Pneumatic	
Motorised	
Gate	
Butterfly	
Safety	

Gauges

General

Thermal conductivity

Hot cathode ion

Hot cathode BA

ISO	DIN
Cold cathode ion	
Partial pressure	
Differential membrane	

Other components

Baffle or trap

Baffle with cooling

Cold trap

Cooled baffle

Bell jar

Symbols to
ISO 3753 - 77
BS 5543 - 78

DIN 28 401

List of Symbols Used in the Text

A	area
B	magnetic flux density
C	conductance
C_L	conductance of a long pipe
C_0	conductance of an opening
d	molecular diameter
D	diameter of a pipe or circular opening
e	electronic charge
h	height
H_A	molar heat of adsorption
i, I	current
I^+	ion current
I^-	electron current
J	flux of molecules per cm^2 per second
J_C	condensation flux
J_E	evaporation flux
k	Boltzmann's constant, general constant of proportionality
K	gauge sensitivity
Kn	Knudsen number
l	electron path length
L	pipe length
m	mass of a molecule
M	molar mass (gram molecular weight), relative molecular mass (table 1.2)
n	neutron

n	number density of molecules
n_M	number of moles
N	number of molecules
N_A	Avogadro's number
p^+	proton
p	pressure
p_i	partial pressure of i th gas
p_e	equilibrium vapour pressure
p_u	ultimate pressure
P_0	atmospheric pressure
q	binding energy of adsorbed molecule
q_G	specific gassing rate
Q	throughput
Q_G	gas load due to outgassing
Q_L	leak rate
Q_P	gas load due to process-created gas
Q_T	total gas load
Q_V	gas load due to vaporisation
r_0	average surface-adsorbed molecule distance
R	universal gas constant, radius
RH	relative humidity
Re	Reynolds' number
S	pumping speed (volumetric flow rate)
$S*$	pumping speed at pump inlet
t	time
T	time constant

T	absolute temperature
T_{c}	critical temperature
u	average drift velocity of a molecule
U	imposed velocity, potential
v	volume of a chamber being evacuated, velocity of a molecule
$\bar{\mathrm{v}}$	average velocity of a molecule
$\overline{\mathrm{v}^2}$	mean square veloctity of molecules
V	volume of a chamber
V	volume of gas
$\dot{V}$	volumetric flow rate of gas
W	mass of gas
$\dot{W}$	mass flow rate
Z	atomic number

α	transmission probability
η	viscosity coefficient
κ	thermal conductivity
λ	mean free path
ρ	density
σ	ionisation cross-section, tangential momentum accommodation coefficient
τ	stay time
ϕ	potential
ω	angular velocity
$\dot{\omega}$	angular deceleration/acceleration

List of Units Used in the Text

amu	atomic mass unit
A	ampere
Å	Ångström unit (=10^{-10} m)
cm	centimetre
eV	electron volt
g	gram
h	hour
kg	kilogram
K	degree Kelvin
l	litre (=10^3 cm^3 = 10^{-3} m^3)
l s^{-1}	litre per second
l min^{-1}	litre per minute
m	metre
mm	millimetre
m^3 h^{-1}	cubic metre per hour
mbar	millibar
mbar l	millibar-litre
mbar l s^{-1}	millibar-litre per second
min	minute
Pa	Pascal (=1 N m^{-2})
s	second
slm, sccm	standard litre, standard cc, per minute
T	Tesla
V	volt

Index

For Product Safety Concerns and Information please contact our EU
representative GPSR@taylorandfrancis.com
Taylor & Francis Verlag GmbH, Kaufingerstraße 24, 80331 München, Germany